t² SOLUTIONS LTD

t² Solutions are leaders in the development, marketing and support of Computer Aided Design (CAD) solutions for the building industry. Over the past ten years their CAD systems, originally the renowned Rucaps, and now Sonata, have earned a reputation for excellence with architects, engineers retailers and facility managers alike.

t²'s CAD systems have their roots in research carried out in the early 1970s at Liverpool University's centre for Computer Aided Building Design. Here the founders of t², John Davison and John Watts, helped develop a concept unique to computer aided design – that of building modelling.

In 1972 John Davison joined the GMW Partnership who fostered the development of this unique concept. In 1976 the resulting system called Rucaps was ready for its first project – the massive King Fahad University in Riyadh, Saudi Arabia. Rucaps proved so effective in co-ordinating all the production drawings for such a complex project that in 1977 GMW helped set up a separate company called GMWComputers, to develop and market the system to other design practices.

During the next ten years, Rucaps underwent significant enhancement to progressively meet the needs of not only architects, but also other designers and users of buildings such as services engineers, retailers and facility managers. World wide distribution and its expansion beyond the building design profession resulted in the Company changing its name in 1987 to t² **Solutions**, the t² symbolising the product of two key ingredients of its success – **teamwork and technology**.

In the same year t² launched a low-cost micro CAD system to complement the higher powered mini-based Rucaps. This was followed in 1988 with the introduction of a revolutionary new CAD system called Sonata whose ease of use, functionality, state-of-the-art graphics and high performance heralded the arrival of the second generation of building modelling CAD.

As the business grew in the early 1980s t² observed that its most successful clients had reshaped their organisations to exploit the new technology that they were buying. t² therefore teamed up with a long standing client – BCCH Ltd, a Hertfordshire based firm of architects, to provide CAD management and training services to help all users. Known as **T Three** (Thought Technology Technique Limited), the joint venture brings advanced project management techniques to bear upon clients' organisations, their projects and their CAD systems.

t²'s continuing mission is to provide state-of-the-art tools built with a specialist knowledge of project management techniques to satisfy all the needs of a building's designer, its builder and its owner.

The Management of CAD for Construction

STANLEY PORT

Sponsored by t² *Solutions Ltd*

BSP PROFESSIONAL BOOKS

OXFORD LONDON EDINBURGH

BOSTON MELBOURNE

First published 1989

British Library
Cataloguing in Publication Data
Port, Stanley
  The management of CAD for construction.
  1. Civil engineering structures. Design
  applications of computer systems.
  Management aspects
  I. Title
  624.1′771′0285

ISBN 0–632–02055–5

BSP Professional Books
A division of Blackwell Scientific
  Publications Ltd
Editorial Offices:
Osney Mead, Oxford OX2 0EL
  (Orders: Tel. 0865 240201)
8 John Street, London WC1N 2ES
23 Ainslie Place, Edinburgh EH3 6AJ
3 Cambridge Center, Suite 208, Cambridge,
  MA 02142, USA
107 Barry Street, Carlton, Victoria 3053,
  Australia

Typesetting by DMD, 52 St Clements, Oxford
Printed and bound in Great Britain by Redwood
Burn Ltd

Contents

Preface

In the era of Information Technology, the computer is the machine-tool. Designers and planners are information workers and many have turned to CAD technology, hoping to find something that will ensure survival in the increasingly competitive business climate. The new problem relates not to any limitations of systems, but to the lack of knowledge on how to implement, manage and control the CAD technology.

This book is aimed at design professionals, planners and managers. Although references and examples relate to building and construction work, most of the principles are unlikely to differ whatever the application. As a result, it should be useful in the fields of mechanical engineering and manufacturing industry too. Chapter 13 deals with applications in construction planning, space planning and facilities management. Emphasis throughout is on people, responsibilities, applications, organisation and procedures.

The design process is highly interactive. Manual drawing, or use of a computer drafting system to mimic this, inevitably leads to inconsistencies within in the design information. Computer modelling of projects presents better opportunities and the many techniques range from 2–D modelling to solid modelling. A blend of 2–D and 3–D methods to suit the application is essential today.

System planning itself requires a carefully managed feasibility study comprising preliminary and detailed phases. Objectives and requirements of the office must be set down. Then there is something to compare the available systems with. The chosen system must be capable of evolving to meet an ever-changing future.

Every CAD system requires a CAD Director, a CAD Co-ordinator to concentrate on the applications, and a Computer Manager to look after the system components. These are management roles; they do not necessarily have to be performed by three different people. The important thing is that the various responsibilities are defined and allocated. Operators have to be selected with care, and training extends far beyond the initial training of operators. Organisation of work-sessions, accommodation for equipment and work-terminals, procedures for security and maintenance are all required. Every system needs objectives, and the monitoring to ensure that these are being met.

Preliminary design of projects can involve creating 3–D prototype models. Then schemes can be assessed and presented visually. Detailed design cannot in practice adopt 3–D modelling throughout, but methods which bind 2–D component representations with their 3–D locations have potential. Design offices must seek to accumulate and manage their design information – graphic and text data – so that it is reusable in future projects.

It is essential to select suitable projects for CAD application. Project management involves predicting the CAD resources needed, and the planning of tasks and required documents. Project control procedures can include the monitoring of the document production, the important phases of checking, approval and release of information, together with document issue, revision control and audit generally.

Acknowledgements

First I must acknowledge the considerable assistance which I have received from t^2 Solutions Ltd. This company has not only commissioned the writing of the book, but has handled many of the administrative arrangements with the publishers. As an independent consultant, t^2 Solutions has given me complete freedom in my endeavours, applying absolutely no pressure or restrictions either on me or on the contents of the book. In particular, I wish to thank John Watts of t^2 Solutions, who has personally provided so much assistance and encouragement. Without this help from John and t^2 Solutions, I feel sure that this book would never have materialised.

I am also indebted to members of T Three – Thought Technology Technique Ltd, a sister company to t^2 Solutions. Colin Connelly has personally provided invaluable assistance in the form of ideas, checking and constructive criticism of the manuscript.

CAD users in various parts of the world have contributed examples of their work which I have been able to use as illustrations within the book. Their generosity has been acknowledged where their work appears. I believe the results clearly demonstrate the high standards that are now being routinely achieved in their offices with CAD.

Stanley Port
44 Busbridge Lane
Godalming
Surrey GU7 1QD

Foreword
by John Watts
t^2 Solutions Ltd

Today's user of Computer Aided Design (CAD) benefits from a decade of experience gained by those pioneers who took on this new technology in its first generation. Whilst these users wrestled with unfamiliar technology and brand new working practices, we system developers had to run fast to remain just one step ahead as we learnt and then adapted our software to fulfil the needs and aspirations of these pathfinders.

At t^2 we have now had over ten years' experience of satisfying the voracious demands of the users of these new tools as they have succeeded not only in driving up productivity but also in continuously expanding their application of CAD beyond drawing production into design, analysis and building management. It has been an exciting and rewarding decade that has culminated, for us, in the introduction of our second generation product – Sonata.

Whilst Sonata brings together the very latest in computer technology and makes the very best use of new software tools, its real achievement must be seen in its ease of use and openness to individual tailoring and to linkages with other applications. These aspects form the benchmark to which all future CAD systems must aspire to if the creative user is to continue to make the progress he has achieved thus far.

Is new technology the whole of the answer? One thing we have learned over the years is never to underestimate the resourcefulness of the CAD user as they have often bent the most inflexible of tools to achieve the most amazing results. Indeed the achievement of many micro CAD users frequently puts to shame the results of the most sophisticated main frame or mini computer user who has technology with ten times the power at his disposal. Why should this be? Can it be that newer technology and software has magically solved all the frustrating problems at a stroke?

The truth of the matter is that any tool performs best in the hands of someone who knows how to use it effectively. In the case of CAD tools, management is the key to productivity and thus return on investment. Furthermore, CAD has been proved over and over again to significantly improve profitability despite the prognosis of those who haven't exercised effective management control over their systems.

Recognising, therefore, the need to ensure effective management, t^2, together with BCCH set up a company named T Three to specifically provide guidance and training on how to set up and manage a successful CAD installation. The results of this partnership have been very fruitful with busy companies achieving high levels of productivity and despondent users being turned into profit makers within very short timescales.

Being pioneers ourselves in this exciting industry, we recognise the need to see CAD, of whatever flavour, being universally acknowledged as a desirable investment. Whilst

T Three can help many they cannot reach all CAD users. Therefore we decided to invite a leading, and impartial, authority on CAD in the construction industry to produce a book that would help and assist new and existing users of CAD to employ these tools more effectively and profitably through correctly applied management techniques.

Dr Stanley Port was our first choice for undertaking this task as his previous book *Computer-aided Design for Construction* demonstrated his thorough understanding of the complex web of alternative solutions available in this field and their applicability to the many different facets of the construction industry.

The brief to Dr Port was to produce a book which was of use to the general CAD user and which was not biased towards our particular solution. Similarly whilst encouraging him to talk to T Three and t^2 users in seeking examples and illustrations for his text, we put no restrictions on his research and sources.

The result matches the specification; no one who reads this book, whichever CAD system they're using, will put it down without having discovered some new way of improving their profitability.

CHAPTER 1
Introduction

In the latter half of the 1980s, companies throughout the world are experiencing an increasingly competitive business climate. They have to face this fact squarely and find an intelligent response. The alternative is to ignore it and ultimately suffer a declining workload or worse.

A poignant example from the past can be seen in the UK steel industry. By 1979 this was in serious trouble and was making huge financial losses. Since then, productivity at the main plant which produces structural sections has been increased three-fold[1]. The steel fabricators also responded. They improved their efficiency and productivity by 45% in terms of output per man in a period of just four years. As a consequence, we now see a vast improvement in the health of the steel industry in general, and a resurgence in the use of structural steel in building construction in particular.

It seems clear that the latter part of the 1980s and the 1990s will be looked upon as the era of Information Technology. The computer is of course the 'machine-tool' designed to deal with information in its various forms.

If we look at the role of the computer generally we find that it has already become an essential tool in many walks of life. In the financial services sector or 'city' for instance, the computer has been at the centre of a revolution in working practices. Its usefulness has resulted from its ability to handle information quickly and accurately. Likewise the airline and holiday industries have whole-heartedly embraced the computer. Again it is its information-handling capability that is sought. Essentially of course, their problem is to get the right people on to the right aeroplane at the right time.

The construction industry is hardly a service industry like these examples, but there are surely parallels that can be drawn. In 1987 it was a business valued in the United Kingdom at £23.6 billion, or 5.8% of the nation's GDP. A vast array of men, materials and machines have to be assembled on site to allow any new building job to 'get off the ground'.

Before that can happen, the designers must process a huge amount of information. Indeed designers are 'information-workers'. They gather data on clients' wishes; the needs of the building users; the requirements of the legislators and statutory utilities; available materials: also they apply their own knowledge of good work practices. All such input data is 'processed' by the designers. The output from their efforts is in the form of drawings, specifications and the other documentation necessary for the project to be built.

Space planners and those people charged with the management of the numerous facilities in and around buildings likewise expend much of their time and efforts on the handling of information.

Of course it would be hardly complimentary to describe our designers, planners and managers as mere data processors. There is much creative effort and human judgement involved, and that certainly is not the kind of work that can be done by any computer. So we should not worry about the spectre of automatic methods of design. We must recognise however that information processing is an important element of the work of all of these people. Also, there is now the new machine-tool which can help to mechanise some of this activity. Therefore there is much scope today for computer-aided designing and planning.

Computer technology has certainly been advancing with breathtaking speed during the last few decades. Computer Aided Design (CAD) harnesses the new techniques of interactive computer graphics and attempts to apply them in the service of designers and planners.

In the early days of computing, every user had to be an enthusiast and have a thorough understanding of the *bits* and *bytes*, much in the same way that early motor-car drivers (referred to at one time as 'engineers') had to understand their mechanical gadgets. Today our construction engineer, architect or planner can indulge in long-distance driving, whilst having little knowledge of mechanisms. What is even more important today is the knowledge of what car to buy, how to drive it effectively, and how to find your way about.

It is the senior management people in design organisations that have to take the decisions regarding the implementation of new technology. Unfortunately they do not have many computer tools to help with their own day-to-day activities. As a result, many of them tend to have little first-hand understanding of computers.

Indeed many of them only pay lip-service to the role of the computer. They may be aware of its use elsewhere, and would always acknowledge in public the need for progress. Perhaps they consider that they are too busy with what they regard as their own profession's work to make the effort necessary to come to terms with new computer technology. Others are privately downright sceptical. There may be lingering doubts regarding the cost-effectiveness of computers. There may be worries about the seemingly rapid obsolescence of equipment, long before it has had time to wear out. There may also be a general sense of unease, not so much with the rapidly moving technology itself, but with exactly how to apply it in individual circumstances. Uncertainty breeds fear – a fear of the unknown. No positive decisions are ever made where there is uncertainty.

Even in some offices where CAD is already implemented, there is a feeling that the system is not producing the type or scale of benefits that were originally envisaged. Indeed management sometimes places far too much emphasis on a narrow view of cost-justification. They fail to understand that improvements can be made to their working practices, the designs which they produce can be of higher quality, and by such means their business can be more cost-effective.

When any new technology starts to make an impact, the things that become important are the knowledge of:

(1) The opportunities offered by the technology.
(2) The limitations of the technology.
(3) How to implement, manage and control it to best effect.

Ready-to-use CAD systems can now be purchased from specialist suppliers in forms suitable for almost any kind of organisation and requirement. However, truly practical and economically-usable packages have been available for less than one decade. There is a shortage of real experience and most existing users only have one or two years' experience. Management needs sufficient knowledge in order to know the right questions to ask.

Even when such decisions have been made, management has the tendency to lose interest too quickly. They feel that their job is over once the cheque for a CAD system has been signed, and a few operators have been through a training course. In reality, management's job is only just starting at this point.

The new technology is rarely introduced to green-field situations where a fresh beginning can be made. Instead it must be eased into existing management structures with the minimum of adverse effects. The on-going business must go on. What is needed is evolution in methods, not a revolution, for the latter could so easily cripple a business.

It is certainly now time for design organisations to come to terms with CAD. Knowledge must be acquired; plans must be formulated. These are vital if the money and effort is to be invested in meeting the real needs rather than merely on installing technology for its own sake or to impress others.

What is needed there is:

- A proper business plan to be initiated for the organisation – detailing the overall objectives, management structure and procedures. This will provide a framework into which CAD can be fitted.
- A CAD strategy must be formulated. This requires some CAD knowledge or advice.
- Responsibilities must be defined. In particular somebody will have to function as a guardian of the CAD strategy.

This book has been written in the belief that the problem facing us today is not the lack of suitable technology. Rather, it is the lack of knowledge of how to implement, manage and control it. The book is aimed chiefly at the professionals and managers of planning and design organisations. The intention is to look at the design process, and see how people can organise themselves so that their project work can be accomplished more effectively with the aid of computers. For the most part we shall be discussing ideas and practices which already have currency. Indeed most have already surfaced within the best-managed offices. This makes it possible for the experience of the forerunners to be examined and used profitably elsewhere. There is little need nowadays to leap into the unknown.

The emphasis throughout the book will be on people, applications and procedural matters rather than on equipment or software.

The bulk of the examples and illustrations relate to the construction industry. However, CAD is applied in many industries. The book concentrates on the principles involved, and most of these are likely to be relevant irrespective of the particular application or industry, so the book should be useful, for example, to mechanical engineers too. I make much use of the word 'designer' which is intended to cover the people working within a wide range of professions and disciplines.

CHAPTER 2
Designing For Building Construction

This book investigates the ways in which we can use computers for design and layout planning. There is little prospect of fully-automatic design by computer and perhaps we should be thankful for that. Therefore we will always have to pick and choose areas in which computers will be employed. Conversely, we need to know where to avoid applying them.

It is worth devoting one chapter just to looking at the nature of the traditional or 'manual' design process. In so doing, perhaps we can gain an insight into the strengths and weaknesses in the existing design procedures.

Although design work is highly iterative, we shall examine the various facets. These include the emerging problem expressed in the design brief, the generation of creative design ideas, the 'storage' of these on paper, and the need for their retrieval and communication. The traditional design process is highly dependent on making drawings on paper. I call these *paper models* and will look at their advantages and limitations. Basically the problem is that the design solution is expressed as a series of unconnected 2–dimensional images throughout the process. This results in poor design co-ordination. Too often a drawing has to be scrapped and restarted just to accommodate a few changes. Sometimes resort is made to building a scaled physical model of a project, but this is often adopted only for presentation and public relations reasons.

We seem to need a better design tool to help in project modelling.

2.1 THE MEANING OF DESIGN

What do we mean by the word 'design'?

Design process
Sometimes we use the word to denote a process. This includes determining the needs of a situation, and conceiving a solution. The process also involves analysing and working up detailed proposals, and then the production, and issuing documentation. These are all features of the planning needed for any new project or facility required by a client body.

Design solution
Alternatively we use 'design' to denote the product or solution arising out of this process. Then it is an abstract term used to represent the concept, proposals or layout plans as they exist at any point during the term of the process. Note that the design is not necessarily

the final product of the process; rather it is something which continually evolves until it does become the final or adopted solution. A design solution may exist only in a designer's mind. Alternatively it may be expressed in the form of drawings, specifications and other documents, or in some other type of model of the proposed project.

2.2 REVIEW OF THE DESIGN PROCESS

When we set out to make something which is new, it seems sensible first of all to define the needs of the situation. Then it is necessary to create an answer or solution that seems to fit the perceived requirements in the best possible manner.

Plan of work

The RIBA *Job Book*[2] sets out a framework of procedures for running a building project. This is the 'Plan of work' and the stages outlined are:

A Inception   (Briefing)
B Feasibility
C Outline proposals   (sketch plan)
D Scheme design
E Detail design
F Production information (working drawings, etc.)
G Bill of quantities
H Tender action
J Project planning
K Operations on site
L Completion
M Feedback

This 'Plan of work' applies particularly where design professionals are appointed and a main contractor is chosen as a result of competitive bidding. It is of course the architects' view of design procedure; engineers and others involved in design and layout planning like to take an independent view of design procedure.

Exploration of design problems

By setting out a list of activities like this, the RIBA Plan of Work does tend to imply that the design process can be neatly parcelled into discrete stages, each of which can be dealt with one after the other. Indeed the *Job Book* warns that abortive work is likely if there is a serious departure from the sequence. This could be the result for example if the brief is modified after the scheme design stage is passed. Abortive work could also result if any change in location, size, shape or cost is made after the detail design stage is finished.

Lawson[3] discussed the design process in some depth. He pointed out that any design problem cannot be comprehensively stated. The client's brief is just one expression of the design problem, but the problem itself evolves as solutions are attempted by the designer. The design objectives and priorities change as solutions emerge. Indeed the design process involves identifying and exploring the problem, as well as solving it.

This view is after all in accordance with our experience that the more a design problem is worked at, the more (and hopefully – the better) are the ideas that occur to us. All the time, we come to understand more about the nature of the needs themselves. It is essentially a highly iterative process rather than a linear one.

Lawson also pointed out that the RIBA Plan of Work is a description of the products of a process – the feasibility reports, sketch plans, production drawings – and not a description of the process itself. He refers to it as the architectural profession's propaganda exercise, its view on how the business transaction among the client, architect and other parties should proceed.

However, as a guide to project administration, the Plan of Work can certainly be a helpful framework for the design team.

2.3 DESIGN IDEAS

Design activity seems to be bound up with the continual clarification of the requirements – the problem – and with the generation of creative ideas in the search for a solution. Since the problem cannot be defined comprehensively, Lawson concluded that there is an inexhaustible number of design solutions.

Source of creative ideas

Design solutions must first take shape within the designer's mind. We shall not concern ourselves here with the process by which the designer generates creative ideas. Clearly this is a fundamental element of the design process but it is a functioning of the human brain that remains poorly understood. For the time being at least, it remains essentially a human capability.

With the emergence of artificial intelligence, some believe that computers will soon be capable of generating results that have some similarity to creative ideas by humans. However, the practical designer can afford to regard such notions with scepticism until such time as they can be demonstrated to work. Meanwhile it is clear that some designers are more prolific than others in the generation of creative ideas.

Initial selection

If there is an inexhaustible number of possible design solutions, clearly there must be a mechanism for accepting, modifying or rejecting design ideas as they emerge. We must be selective and then save in some way all the ideas which are judged to be useful.

Initially there is little more than subjective judgement available to measure how well any one design solution matches the requirements. Also, the subjective judgement of different persons – the client, users, observers, critics, constructors or designer himself – will all vary. This is because each perceives different needs and has different priorities. So discussion of what is a 'good' design or a 'bad' design is always controversial.

Nevertheless we can readily see that the designer's aim should be to achieve what is in his opinion the 'best' design possible within the constraints. Stated in fairly imprecise terms, the solution chosen must be the one that appears to meet as many of the main requirements as possible. Another way of looking at it is that it is the solution which is likely to be readily acceptable to most of the key persons. Good design therefore involves human judgement, and selection of ideas is usually done to achieve some sort of a compromise among many widely differing needs.

Of course, by no means all design choices are made solely by subjective means or based purely on past experience. There are all kinds of statutory and other requirements, rules of thumb, standards, codes of practice and accepted conventions.

Logically, the more time available to the designer, the more options that can be generated, assessed and selected. Therefore the designer can always find a way of improving his design.

2.4 STORAGE OF DESIGN IDEAS

Stored design model

If the project is of very limited scope, such as a small do-it-yourself task around the house, then the actual construction work might well proceed straight from the design concept which is held in the mind of the handyman himself. The stored design model is all in his mind.

In most other construction situations however, there are far too many design details involved for these to be retained in any human brain, if accuracy and speed of retrieval is to be guaranteed. A typical construction project may contain many thousands of components. Traditionally, of course, the design model is maintained throughout this evolutionary period on sheets of paper.

Evolution of design models

The dynamic nature of the design must be stressed. It gradually emerges like some crystalline structure forming within a 'soup' of the requirements and constraints of the situation. 'Crystals' form on the germs of ideas. They seem to change shape as they are looked at from different directions. They break down, change, merge and reform. All the time the structure is growing to a higher and hopefully more orderly and useful level of organisation.

Our design model must not only be capable of simply storing the design information. It is imperative that it can permit additions and alterations to be readily made as well. If there is difficulty in absorbing new items then naturally this can have a bearing on what can be achieved by the designer in a finite time. A halt might well be called to the design process before the design has had much of a chance to improve and mature. The quality of the design, and perhaps the success of the project, might be affected to some degree by the ease with which information can be absorbed into the stored design model.

New contributions or design changes

This notion too accords with our experience. However there is a law of diminishing returns which applies here. Each new creative idea must be evaluated, then accepted, modified or rejected. If accepted it has to be integrated somehow into the stored model of the design.

If this act of integration can be done with little time and trouble, then the idea is regarded as a 'new contribution' to the design and it is welcomed.

However a design idea might well be judged as a necessary improvement, yet it is difficult or tedious to incorporate it in the stored design model. This design idea is then labelled a 'design change' or a 'revision'. This is often unwelcome and is potentially costly.

There is therefore no intrinsic difference between the 'new contribution' and the 'design change'. The difference in view depends on the logistics of their incorporation into the stored design model.

Halting the design process

Since the design can always be improved with more effort, the design process in practice proceeds until someone calls an arbitrary halt. This can be when the flow of ideas is no longer producing significant or worthwhile design improvements, or when too many of the ideas are seen to be in the nature of costly 'design changes'.

Of course no designer in the construction industry has the luxury of having unlimited time at his disposal. The client initiates the process and usually imposes time constraints. The fee structure also imposes a practical limit on the manhours that can be spent. So the designer has to do the best he can within these imposed limits.

2.5 RETRIEVAL OF DESIGN IDEAS

Visualisation, analysis, assessment

In addition to the *aide-mémoire* or storage medium for design ideas, the designer also requires a means of retrieving quickly and accurately any aspect of the saved design. He must be able to visualise the design, and this cannot be left entirely to his own powers of imagination. Others too, such as the client, will wish to visualise the proposals. Again, the traditional means of retrieval is by using paper. The designer can search through his papers and can visualise the evolving design by finding or making anything from simple doodles and back-of-the-envelope sketches, to formal project drawings. Occasionally other methods are used, such as building a physical model.

Information may have to be retrieved to permit some type of assessment of a design solution. In some cases it is possible to examine some aspect of a proposed solution using an appropriate analytical procedure. Thus we can perform a structural analysis of a building frame, or make calculations relating to heat losses or gains, acoustic or lighting performance. Such calculations are made in the belief that interpreting the results will provide an insight into the suitability of a proposed design solution.

Cost pressures in design process

Mechanical, electrical and electronic engineers often find themselves designing components which will be manufactured in hundreds or many thousands of units. The cost of design in these circumstances can be a tiny proportion of the manufacturing costs involved in a total production run. Alternatively there may be merit in allowing the design process to extend with the aim of finding a design solution which approaches rather closer to perfection. There is often a justification for building mock-ups and physical prototypes for testing. Adjustments or fine-tuning to the design often can be made during the production cycle of the product.

All this is not commonly found in the construction industry. Perhaps it is only in housing, particularly public housing and low-cost private housing, where similar units are built in large numbers. However, the arrangement of these units within a landscape and in relation to their neighbours still creates unique design problems.

When we look at most other building work, we generally find that each project is unique and so each is separately designed. A project is typically large and complex, and may require several years from inception to completion. The construction has to have a long useful lifespan and so there is stress on durability and safety. This means that construction designers do not normally have the chance of seeing mock-ups or prototypes. Nor do they have much opportunity for fine-tuning the finished project in the light of experience of production or use. In short, project uniqueness means that the design must be right first time. This lays special stress on the designer's ability to visualise and otherwise assess the design solution while it evolves.

All the time, as everyone knows, there are cost pressures on the design process. Design must cost no more than a few per cent of the construction cost of the unique product – not a few per cent of a total cost of a long production run of many similar items.

2.6 COMMUNICATION OF DESIGN IDEAS

Multiple disciplines

We have discussed the role of 'the designer'. However, in reality nearly all projects are of such a scale and complexity that several types of designer are needed. There may be architects, structural engineers, HVAC and electrical engineers, quantity surveyors, drainage experts, landscape architects, surveyors, interior designers and so on. They all have to work in parallel, not sequentially. So they have somehow to keep track of what they are doing as individuals, and as a design team.

These designers must all co-operate in a multi-discipline team. A team is often created specifically for one project; the next project will be handled by another *ad hoc* assembly of designers. Some of the different disciplines may operate under the umbrella of a multi-discipline design practice. More often they each operate from individual single-discipline practices. So we have a very fragmented design industry, composed of a multitude of small units. If the design process itself had to be designed afresh by any of these design disciplines, they would hardly come up with this solution! So there is much need for effective communications within the design process.

Separation of design and construction

Of course, it is only when the design concept has been extracted from the minds of the designers and expressed in some tangible format that construction or fabrication can proceed. It is customary ever since the time of the industrial revolution for the design process to be undertaken by professional designers. These are people not actively involved in organising the construction work. Even in design and construct situations, the designers usually work within a design department, and are not involved directly with site operations. So another reason for needing an explicit description of the design is quite simply to permit construction to be carried out by persons other than the designers.

Communication network

There are many other important communication paths. The client will want to be kept informed regarding the proposals as they evolve.

Abstract concepts are of little help here and most clients prefer visual images or physical models.

The users of the proposed facility may well be a different party from the client. For example the user of a hospital is not the hospital authority; there are many users including the doctors and other medical staff, and the patients. Users of a facility often have different needs, sometimes even conflicting needs, compared with those of the client body. So the users may be another class of interested parties which has to be consulted. Inevitably an expression of the design in a suitable format is required to make this possible.

Among the many other interested parties are utility companies, traffic authorities, planning authorities and the building control authority.

The various design professionals must not fail to attend to the marketing of their practices. This means that selected project information has to be presented to learned societies, the news media and to publications of various kinds. Each recipient has different needs and ideally requires different aspects of the design concept to be expressed or highlighted.

These expressions of the design in tangible forms are therefore required for purposes of communication. Traditionally they are records – usually drawings – on paper.

Design tools

It is clear that the designers are dependent on their own mental abilities for making their assessments of the design requirements. They must generate their own creative design ideas. However, we have seen why construction designers are especially dependent on some form of design tools for the:

storage,
retrieval and manipulation, *and*
communication

of the design model. The quality of the project is in large measure determined by the effectiveness with which these design tools can be handled. Until recently they have been no more that the drawing board, drawing instruments, and sheets of paper. In the next chapter we will examine the role of digital models which can be held in computer memories. Before turning to computer methods however, we must critically examine the advantages and shortcomings of assembling a project design by the traditional means.

2.7 PAPER MODELS

Having worked-up their various ideas into design solutions on paper, the designers can file them away for use later. Sheets of paper can always be available for reference when needed. It is drawings and sketches that have been the favourite mode, for designers understand better than most that when it comes to visualising a project 'a picture is worth a thousand words'.

They do use words to describe design solutions, but mostly when communicating abstract notions like specifying the required practices or the quality of materials to be used.

Advantages of paper

Images drawn on paper to depict aspects of the design are forms of iconic models of the real project. For this purpose, the use of paper or a similar medium has a number of important advantages:

1   Manual drawing procedure is well understood by most practitioners. It has, after all, been developed over hundreds of years.
2   The medium itself is cheap. Likewise the associated tools – pens, pencils, instruments and drawing board are also relatively cheap.
3   Small corrections are easily and quickly carried out.
4   The drawings can be transported, by 'hand' or post. Drawings of limited size can be communicated by facsimile transmission.
5   The drawings can be read by anyone possessing the relevant background knowledge and experience, at any location. No special equipment is needed.
6   Paper (or better still, plastic film) is relatively stable and so can form a permanent record.

Disadvantages of paper

With advantages like these, it is no wonder that paper documents have come to play such a major role throughout society. However, we should not be blind to the disadvantages of drawings on paper.

1 Paper is a two-dimensional medium, and so drawn images can be no more than two-dimensional views of real three-dimensional objects.

2 Drawings can contain little in the way of definitive 'structure' or 'intelligence'. This means that the drawing consists of lines and text characters only. Clearly an experienced person can read the drawing and interpret the meaning of the lines and text. For example the full relevance of a particular graphic symbol may be appreciated. However, many problems occur in practice because of differences in interpretation between the producers and the users of drawings.

3 The size of the sheets which we can handle conveniently is limited. This places an upper limit on the quantity of information that can be placed on a single sheet.

4 A corollary of the above point is that most building projects require many drawing sheets for their description. The more sheets needed, the more cross-referencing that is involved. On very large and complex projects the handling of many sheets becomes almost unmanageable. Cross-referencing is very inconvenient and is the source of many errors.

5 Some aspects of drawing production are very repetitive and boring.

6 Manually-produced construction drawings are usually inaccurate to some degree, and often incomplete. This leads much later to queries from site staff. These cause interruptions to the current work in the design office because by then the site queries tend to be very urgent.

7 A drawing on paper is 'hard' information. While small corrections or amendments can be made easily, major modifications usually involve either substantial or complete redrawing of one or several sheets.

8 Usually the sketch plans are discarded completely and new drawings are produced for detail design. Similarly it is not always possible for the drawings used for tendering purposes to evolve into the construction drawings.

9 When a set of drawings is planned, the scales and contents of each sheet are determined at the outset. By the time the drawings are finished, more is known about how they *should* have been presented. By then it is too late.

10 Control of amendments and revisions is not easy in practice.

Some of the above limitations should not be underestimated, and indeed some of the major problems will now be highlighted further.

2–D images of 3–D projects

Architects and engineers design real three-dimensional projects. It is a severe limitation indeed to have to represent such projects throughout the design process with two-dimensional diagrams only.

There are means of coping with the situation, such as the production of standard projections like plans, elevations and sections. However, in building design the elevations and sections produced are rarely true graphical projections constructed directly from the plan view. They are in many cases entirely distinct two-dimensional views, often incorporating information that is not represented in the plan. Perspective views which could be very expressive and useful are rarely seen, simply because they are so difficult and expensive to produce.

Of course we would be able to understand the proposed design solution better if we could view it somehow in 3–D. However, this is not the basic problem. Of more significance is the fact that planning with the aid of a series of 2–D images inevitably results in poor design *co-ordination*.

As is well known, poor co-ordination is the source of incalculable effects – the errors, costly delays and abortive work that are endemic within this industry. The situation is exacerbated because many of these problems do not come to light during the design cycle itself, but only after the construction is advanced on site. The later the design errors come to light, the more expensive are their repair. In a £23.6 billion UK construction industry, poor co-ordination is indeed an expensive problem. This is not helped by the gulf between the designer and the contractor. In the final analysis, too often it is for the contractor to co-ordinate the work; there are few rewards within the design team for better co-ordination.

Scrap everything!

It is certainly easy to make small alterations or 'new contributions' to a drawing during its initial preparation. Major design changes are another matter. Some of these involve the redrawing of many sheets from the beginning, and we know how boring, costly and prone to errors that this can be. The special difficulty to be faced is that any one designer has difficulty in being absolutely sure that all the knock-on effects of his own changes have been dealt with and checked.

Once a drawing has been finished and issued however, any subsequent design changes are far more likely to be costly and unwelcome. This is particularly true in the multi-discipline environment. Formal drawing revisions have to be invoked and reissues to all affected parties have to be made. Everyone has to check on the effect of the changes on their own work. Slow-speed postal communications with paper drawings often means in practice that a designer is working with out-of-date copies of drawings issued by other members of the team.

Designers have to struggle with huge numbers of original documents, with multiple copies needed of each of these. The cost and inconvenience of storing them, of retrieving information from them and of exchanging them with other parties is all far from negligible. The size of plan chests and the difficulty of finding the needed drawings in the average drawing office is a reminder of this.

2.8 SCALED PHYSICAL MODELS

In many situations, paper drawings are judged to be so inadequate for the needs, that they have to be supplemented by scaled physical models. This practice is common in other industries. For example, clay models are made for new designs of production motor-cars and mock-ups of aircraft cockpits are made to help in the design of the controls and layout of instruments. In the design of chemical engineering and process plants, the complexity in the layout – of pipes, vessels, valves, pumps, supporting structure, cable trays, controls and instrumentation – is such that two-dimensional drawings alone are certainly inadequate. The building of plastic models of such plants has been standard practice. Other examples of the use of physical models are to be found in electronics design – for printed circuit board layout design – and in chemical and pharmaceutical research.

To reflect the third dimension – and for public relations

In most of these examples, the physical models are constructed so as to help the designers involved to cope with the third dimension. It is here that the drawings done on the drawing board are so inadequate.

Of course architects and engineers in the construction industry sometimes resort to making physical models too. They are made usually for large or prestige projects. The prime purpose is often for public relations – to impress the client, users or public generally. This is a reflection again on the inability of drawings – in the form of floor plans, sections or elevations – to adequately communicate an impression of the design to lay-people like the client or public.

The problem with physical models however lies in the high cost of their production. It is not easy to make models of free-format objects. It may be impossible to plan at an adequate level of detail with their aid. Also it is difficult to alter physical models once they are made.

In some industries, the high cost of prototype models is warranted because of:

(a)   Complexity and high value of the project, *or*
(b)   Economies of scale, where repeated use of the design – as in motor-car design – will cover the high cost of the model.

In the building industry, physical models are rarely made until the design has more or less settled down or perhaps not even until it is finalised. They are mostly used for public relations rather than as an aid to the design process.

2.9 THE NEED FOR BETTER DESIGN MODELS

In the construction industry we have therefore a situation where multi-discipline design teams are charged with the creation of design solutions for unique products. There are always many parties with varying interests in the outcome of the process. Nobody will have any chance to see or experience the real end-product – which is always large, imposing and costly – perhaps for years ahead. In such circumstances it certainly behoves the designers to continually review their methods and adopt the best design tools available. Of course these must be useful, cost-effective, and must facilitate communications.

CAD is making inroads in many other industries. We must take a close look at how prototypes of our end-products can be constructed first within computers, before the real projects emerge on site.

CHAPTER 3
Computer Models

In this chapter we turn our attention to the various ways in which the design solution can be represented or modelled within a computer.

First we look at the basic attributes of the computer which makes it potentially useful to the designer. Computer models are created by interaction of a user with suitable programs. The latter, together with a suitable internal data structure for design information, form what we call Computer Aided Design (CAD) systems.

One approach is merely to mimic drawing board practice by 'drawing' the lines and text into the computer. This is computer drafting. Computer modelling relates not to such lines and text, but is more project orientated. Building components of various types are created and assembled to form a computer-based design solution.

Computer modelling can be in 2–D format but we tend to have to break the project into many sub-models – so problems of design co-ordination are not eliminated. Components, represented by 2–D images, can be built into a spatial locational model and this is a stepping stone to 3–D concepts. Other 3–D techniques with increasing levels of completeness of description are discussed. Some 3–D approaches that have been adopted and are better suited for use in other industries. For construction design and layout planning, there is no single ideal method. Rather the designer needs a toolkit of different 2–D and 3–D methods, so that he can apply the correct one to suit the application in hand.

3.1 COMPUTER MODEL: A REPRESENTATION OF A DESIGN SOLUTION

The computer model is a representation of a design solution. It is in digital form so that it can be held and manipulated within a computer.

Such models are normally created by the interaction of a computer operator with suitable programs. These programs are required to enable computer models to be modified, evaluated, analysed, visualised, saved for subsequent use, or transferred for use by someone else.

3.2 THE POTENTIAL OF THE COMPUTER

Computer memory is now a relatively cheap commodity, so the modern computer is capable of absorbing huge quantities of information. This can be stored accurately, and retained more or less permanently.

Retrieval of information from the computer is also easy, provided the user can specify some logical basis for such retrieval. Thus every word of a book could be printed out sequentially with confidence that it will be 'word-perfect'.

The computer can do nothing unless it is instructed or programmed by a human. In general, the writing of such programs is laborious and time-consuming. However, once some logical process has been specified as a program then the computer can faithfully and precisely follow the same process whenever, and as often as, required. Computers are very good therefore at repetition – going through the same process again and again, albeit operating with different information.

Provided somebody has understood the process and has written a suitable program, there is no need for every user of the program to understand every facet of the process. Nevertheless, subsequent users do need to have confidence in the program.

Computers work 'like lightning', for indeed they do work at the speed of light. However, many of their internal operations are iterative. Indeed they can be so very repetitive that in practice a process can sometimes take a significant time.

Given a supply of electricity and a reasonable environment, the modern computer can be worked long and hard. If necessary it can cope with a 24-hour day and a seven-day week, without feeling ill-effects or even getting bored with all the repetition! Of course a little preventive maintenance is needed and very occasionally there may be a breakdown. Generally however, the modern computer has a robust 'health' record.

Of course the designer cannot perform like this. He needs regular periods of rest to recuperate and likes variety rather than boring or repetitive work. Luckily at any time he can switch on or off his modern slave – the computer.

It better to disregard the mystique that has been built up around the computer. Think of it as merely a machine-tool. It is certainly a very versatile and multi-purpose machine. We shall be concentrating on how this machine can can be used to mechanise a part of what has hitherto been a completely manual process.

3.3 INTERACTIVE COMPUTING

Computers have long been able to cope with numbers and to perform complex calculations. Many of these uses are routine and repetitive, like payroll calculations or a structural analysis. After a process using a suitable program has been initiated, we can in those circumstances ignore the machine until it eventually produces its results for us.

Likewise computers can handle huge volumes of text characters. We know that word-processing programs can help us to create and to manipulate a wide range of documents, including the specifications for a project.

For CAD work, computers have had to be programmed to handle geometry, graphical information and images. For input of such information, we require a device such as a data tablet or a mouse in addition to a keyboard, because there is a continual need to indicate positional data to the computer.

Much CAD operation is iterative but not routine. In practice it is best to tackle the work as a series of fairly basic steps. Each step involves human intervention in the form of the issue of a command or order to the computer to do something. The computer responds almost immediately by displaying the results of our command on a graphical screen. Having completed one step, we can then make the next appropriate step. If the result of one of our commands is not to our liking, we can simply cancel its effect and do something else instead.

At each of these steps, we are therefore initiating one subprogram and then supplying the necessary input data or 'parameters'. Subprograms must be available for a range of basic operations like drawing a line, placing a symbol, creating a planning grid, or initiating the plotting of a drawing. This step-by-step procedure is *interactive computer graphics*, for we are continually interacting with the computer as we strive to achieve some useful objective. By running through a lengthy and unique sequence of such basic operations, we can describe a unique design to the computer. Because the sequence of operations is determined by us alone, we can remain in full control throughout the whole process.

CAD software packages

A CAD system is a computer graphics system which enables us to input, manipulate, store, edit, retrieve and output the geometry, graphical symbols and other descriptive material which together represent the real design.

A CAD software package therefore includes a battery of subprograms, plus a data structure or framework which allows the build-up and retention of all the data needed to describe the geometry and other descriptive material of a design.

All proprietary CAD systems differ from one another. First of all, the individual subprograms or commands differ from system to system. Of more fundamental importance, though, there is no accepted or standardised way in which 'external' design information is represented by 'internal' data held inside the computer.

This matter might appear to be academic, but it is of considerable importance to the designer. Some CAD systems are able to hold a representation of two-dimensional graphical images or line diagrams, including the text annotation normally found on project drawings. The object here is to reproduce the traditional project drawings using a computer. These are computer drafting systems. Other CAD systems concentrate more on holding a representation of the form and location of real objects – these are modelling systems. Although all systems differ, we shall now examine some of the main approaches, starting with drafting systems.

3.4 DRAWING BY MACHINE

Basic elements of a drawing

If we examine any drawing (for example Fig. 3.1), our natural tendency is to read from it the message which the draftsperson has been trying to convey. It might be the layout of a building, the arrangement of some components or some construction details. If we examine the actual marks scribed on the drawing sheet we can see that it is composed for the most part of many occurrences of a few basic elements like straight lines, arcs and text characters. Symbols, a drawing border and title block would be composed using the same basic elements.

Each line on the drawing could be represented within the computer with the two co-ordinates of each end-point, plus information giving the line style, line width and colour.

If a subprogram is available in the CAD system to input one line, then the same subprogram will be able to accept every one of the other lines. The procedure is similar although the parameters are different. The computer is very suitable for repetitive operations. The whole drawing can be represented as a composition of many basic elements which are individually stored within the computer.

Benefits of computer drafting

It is at this stage that we might begin to question the point of the exercise. This is just creating the drawing, line by line, using a computer rather than a drawing-board and drawing instruments. To view the drawing we now need a graphics screen. Otherwise we need the plotted product, and this is very similar to the manually-produced drawing.

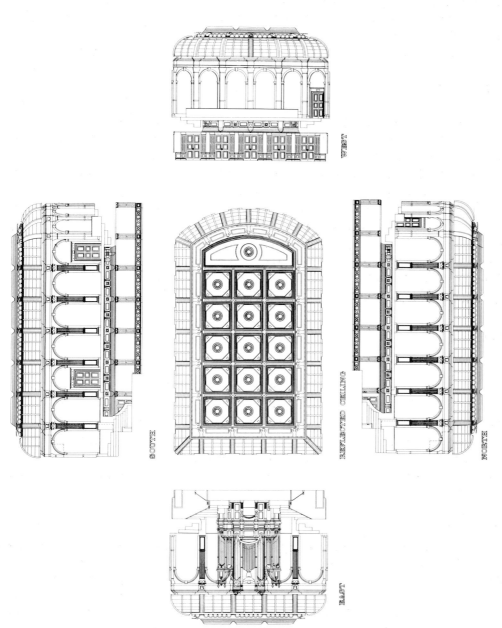

Fig 3.1 Reading Town Hall. (*Courtesy:* Architects Design Partnership, Henley on Thames, England.)

Many would argue that the computer-plotted product is more uniform and legible. Legibility surely helps the communication process which is the basic reason for creating drawings in the first place. A few people, perhaps those who regard the drawing as an end-product – as a work of art rather than as a means of communication – might argue that the machine-plotted product is bland or lacking in character.

The chief aim of drafting by computer is usually to increase the draftsperson's productivity. Whether or not this aim can be achieved depends on many factors. These include the nature of drawings, the skill of the operator, the availability and quality of the design information, and the quality of management applied in the work. These are all in addition to the capabilities of the machine – the computer hardware and software – that has been employed. A productivity improvement factor of three is often quoted as being feasible, but it would be unwise to assume that every system can deliver this. Sometimes this may be exceeded, sometimes it could not. What is certain is that productivity will vary widely over any period.

It probably takes just as long to draw a single line with a computer drafting system as it would to draw the equivalent line with drawing instruments on a drawing board. So if every line and text character of a drawing had to be entered by the operator singly, there would be little or no immediate benefit in using a computer system. Improved productivity flows when:

(a)   Plenty of opportunities can be found for exploiting the computer's ability to repeat similar operations many times over.

(b)   Use is made of the computer's almost instant ability to replicate groups of drawing elements.

At first sight when we look at a typical construction drawing, it might be supposed that few of such opportunities exist. However let us look at some examples of what can be done.

The computer drafting system might well be able to generate very rapidly indeed an orthogonal planning grid of lines for a building. With some systems, this can be done with the issue of just one simple command. The end result can be a grid which is at least as accurate as can be achieved by very meticulous work done on the drawing board.

Either a symbol or the plan view of a building component might be built up from several lines and/or text characters. Created once, the assembly can then be saved within the computer. Then it is available for replicating many times over for use on the same or any subsequent drawing. Thus the plan view of one hotel suite might be generated and then reused as many times as necessary within the same project. The service core of a building may be drawn, then copied – perhaps with some minor changes incorporated – and used on all the drawings depicting the various levels within the building. There is much repetition evident in the drawing of Reading Town Hall (Fig. 3.1), and in the site plans in Figs 3.2(a) and 3.2(b).

The extent to which these and other opportunities can be exploited depends partly on the sophistication of the software package and partly on the skill and foresight of the operator. Of greater significance, however, is its dependence on project knowledge and on the quality of the planning of the drawing work.

Productivity improvement with drafting systems depends too on the computer's ability to process commands rapidly. Preferably it should do this as fast as the operator is able to issue them. Cheap or poorly-developed drafting systems are deficient in this respect. If the operator's thought processes and natural work-rhythm are broken by long waits for the computer to respond, then not only does productivity suffer, but so does the operator's enthusiasm for the new tool.

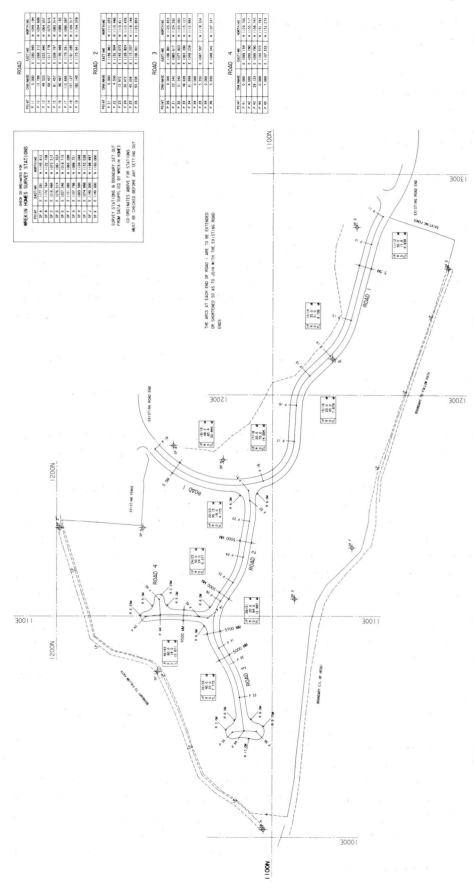

Fig 3.2(a) Site plan showing road layout. (*Courtesy:* Diamond Partnership, London and Wrekin Homes Ltd.)

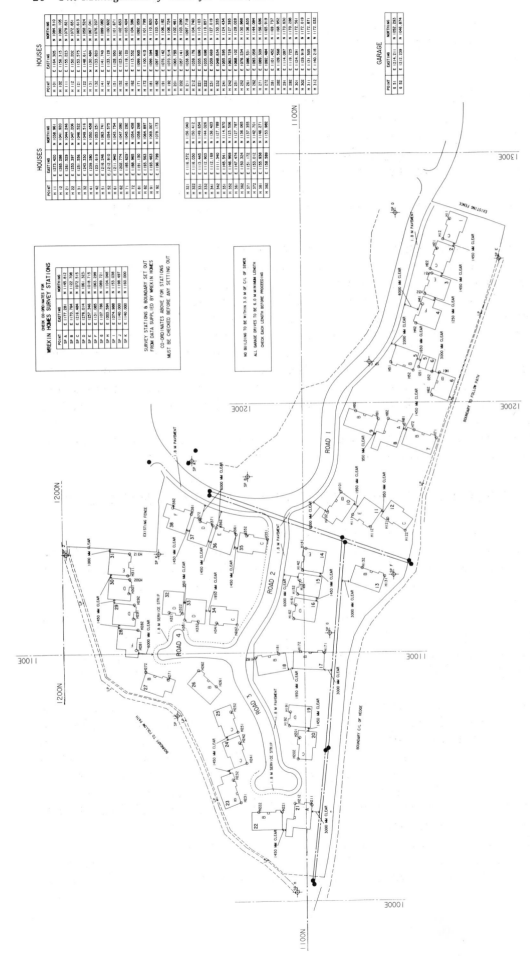

Fig 3.2(b) Site plan showing housing layout. (*Courtesy:* Diamond Partnership, London and Wrekin Homes Ltd.)

The cost of computation falls year by year. There seems to be no end to this process. Likewise as time passes, the cost of professional labour continues to increase. A result is that it is becoming easier and indeed more attractive to implement machine aids which have adequate capacity and power.

Other potential attractions of computer drafting include:

- A reduced elapse time for producing a batch of drawings.
- The probability of rather fewer errors.
- A more uniform quality of drawing.
- Major drawing revisions are undertaken more rapidly, and the new plots are 'clean' – in the sense that there is no evidence of messy alterations.
- The easier implementation of a house-style for a company's drawing work.

The unit of production remains the drawing sheet

So far in this section, our attention has been focussed firmly on the drawing. The aim has been limited to creating electronically the digital equivalent of the paper drawing. This is all that many of today's computer drafting systems can be used for – particularly the low-cost systems. The unit of production of the design office remains the drawing sheet, just as it was in the days before computer systems were implemented.

Any new technology often mimics the old practices, at least when it is first introduced. This helps to limit the cultural shock. We can see examples of this elsewhere, as diverse as the horseless carriage, and the pen-plotter. Later the old and new methods tend to diverge. So to follow these examples, we have the modern motor-car and the electrostatic plotter.

Initially many design offices have introduced computer drafting. Now the advantages of computer modelling are becoming more apparent.

3.5 MODELLING OF PROJECT DESIGNS IN TWO-DIMENSIONS

Think of the project rather than the drawing

A conventional drawing is merely a diagrammatic or symbolic representation of some aspect of a design solution. The draftsperson is always limited by what he can present legibly within the confines of a sheet of paper of standard size.

Computer drafting mimics this form of representation, with all the advantages and disadvantages of computer use superimposed. Drawing information is built up with the help of the computer in the form of lines, arcs and text characters. 'Twist', shown in Fig. 3.3, was a winning entry in the 1987 t^2 Drawing of the Year competition. It is a superb example of what can be achieved with 2–D computer drawing methods – a 3–D system was not used.

Computer modelling by contrast requires a change in attitude. No longer are we just trying to draw with the computer all the lines and text that would otherwise be placed on a drawing sheet. Here the intention is to assemble a model of the project. The aim is to draw the location and details of each and every component or building element which occurs within the whole extent of the project. We rely on the computer system to cope with the large quantity and sometimes high density of the design information drawn. How all this information will eventually be presented in legible form on individual sheets of paper is a separate problem.

As I am referring to 2–D modelling here, the process in practice usually relates to the build-up of a plan model of a whole project site. It might be a floor layout plan of a building. Perhaps each construction component is represented by a suitable symbol.

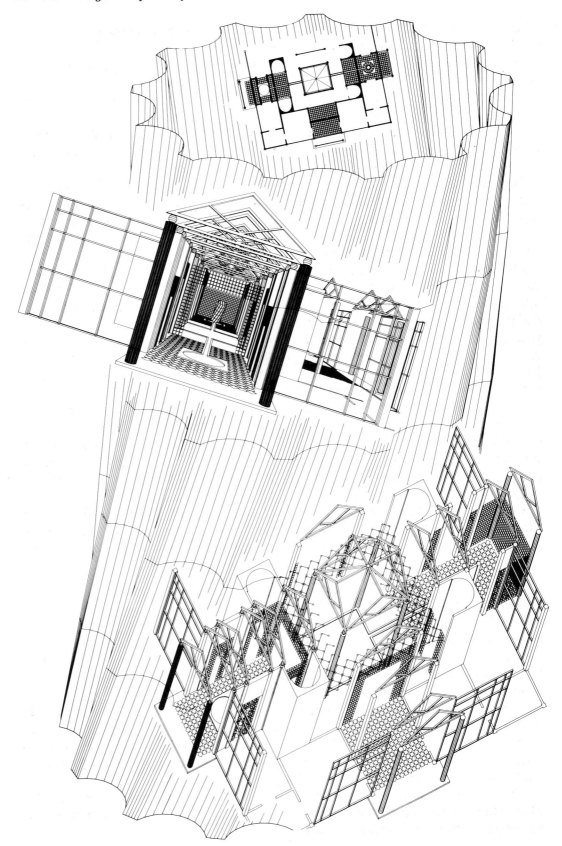

Fig 3.3 Twist. (*Courtesy*: BCCH Ltd and Real Image, Welwyn Garden City, England.)

Alternatively it might be represented by a plan view which shows its outlines, and maybe with appropriate detail and annotation added.

In modelling, the main emphasis is on images which are representations of real components. These representations have to be created, saved, placed within a model, manipulated and classified. The emphasis is not on drawing basic graphical elements like lines, arcs and text characters.

Firstly, each component representation must be recognisable. Second, it must be placed within the model to show the location of the real object within the project. Clashing of any two components should show up visually on the screen.

Computer modelling must be an integral part of the design process itself, rather than being merely a part of the design documentation activity. The evolving design solution is represented at all times by the evolving computer model. The computer can store the evolving design solution, but in addition it can help us to visualise it and to communicate our design proposals at any time to other parties.

Ideally we would like to build just one design model. Then there would never be any doubt as to where elements of the design are located. When we want to retrieve information or need to visualise the design, we would be sure of accessing the current design, rather than out-of-date material. Furthermore with one model we could guarantee that there are no inconsistencies. By contrast, inconsistencies tend to proliferate either when manual or computer drafting methods are adopted. This is because for example a design change might be marked on one drawing but inadvertently not on another containing overlapping information.

A typical design solution does contain a great deal of information, particularly when the detail design stage is well advanced. The simplest computer drafting systems tend to struggle when dealing with the contents of just one information-dense drawing. We can see that model-building presents a much tougher problem in terms of computer processing power and storage capacity needed. The computer system must be up to the job.

Some of the more sophisticated computer drafting systems available today do have adequate power, storage capacity and other features available. This means that they can be capable of working in this two-dimensional modelling mode, at least for projects of somewhat limited scope and complexity.

Definition of drawings and sheets

Inevitably it will be some time before we can have a 'paperless' design process, because too few people have access to suitable computer display screens. So facilities are required in a modelling system to permit the individual views or drawings to be defined. Drawings are usually defined by specifying an area within the model and a suitable drawing scale, and by quoting the classes of design information which are to appear. One or more of such layout drawings, together with a suitable frame and title block may be composed to fit on a standard sheet. With the issue perhaps of a single command, an up-to-date version of each sheet can be plotted out automatically whenever required. By making sensible choices of parameters including the information classes, legible sheets can be produced on any standard size of paper.

It helps if the parameters which define each sheet can be stored within the computer. Then, after the design has been updated, all the sheets needed to depict the revision can be ordered with the minimum of effort.

Sub-model for each level

One problem will be obvious. How can a two-dimensional model represent a multi-storey building? The answer of course is that one model cannot do this. We would actually have to build a model – let us call it a *sub-model* – to depict each significant level in the building. This probably means one sub-model to define each floor level. In addition, ceiling-level sub-models might be needed, especially when multiple services are to be provided above suspended ceilings.

The problem might not end there. So far I have referred to models which produce layout plan views of projects. If the design documentation must also show either elevations or vertical sections cut through our design, then each of these planar surfaces would need its own special sub-model. The computer cannot automatically derive section views and elevations from 2–D floor or ceiling models because it lacks enough information. So these sub-models have to be built up by the designer using elevational or sectional representations of individual components.

This need for many sub-models means that a 2–D CAD modelling system is in practice something of a limitation at least in design. The sub-models inevitably overlap (Fig. 3.4). The same building elements reoccur in more than one sub-model. For example the columns will show up both in floor plans and vertical sections through the building. This means that the same column must be input separately by the operator into each sub-model. Necessary duplication of information like this means extra work and it opens the door to inconsistencies with the design description.

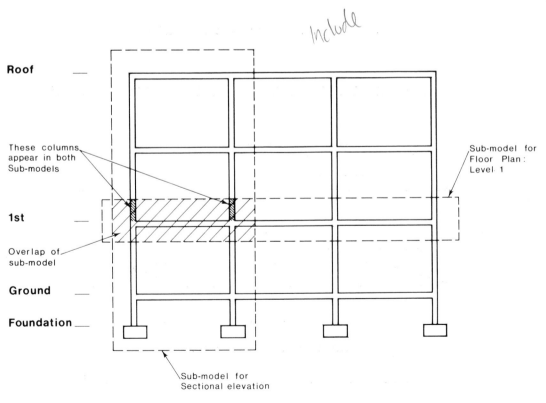

Fig 3.4 Overlapping 2–D sub-models.

3.6 A 3–D LOCATIONAL MODEL WHICH CONTAINS 2–D COMPONENT IMAGES

Layout plans at any level

Drawings may be built up from lines and text characters, but real construction projects are built from real building components and products, and from *in situ* construction. Let us now move gently from 2–D representation towards forms of representation of the 3–D world.

A potentially useful modelling approach is to mimic such construction by allowing each type of component or element in the real design solution to be represented by a 2–D image. Fig. 3.5 shows how a window might be represented. For the moment, I have chosen to represent it by a simple sectional-plan image. This could be 'drawn' or input to the computer once and the image stored there for future use. So far this is no different to the 2–D modelling which I have just discussed.

However, now we could specify where a window occurs in 3–D space within the project. The location could be defined by indicating two plan co-ordinates, plus a level, for the origin point of the image. When appropriate the image of the window could be rotated in plan about its origin point before placement. Having placed one window, we could repeat this process for every window in the building (Fig. 3.6).

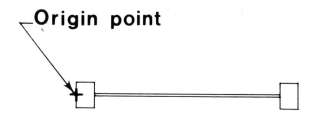

Fig 3.5 Plan image of a component (a window).

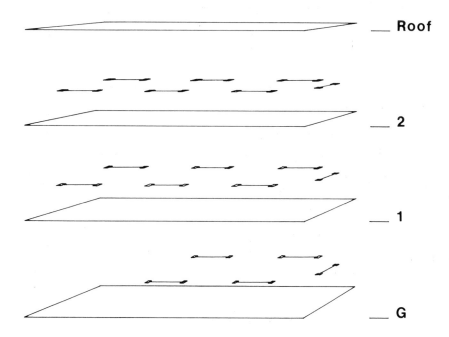

Fig 3.6 Placement of components to form a model.

At this stage, we have a single model which holds the 3–D location of every window throughout the whole building. Also, the computer holds a single 2–D image of the typical window.

With this type of modelling system it is possible now to extract a floor plan for *any* level in the building. Indeed, more usefully, we could specify a range of levels, say from second floor level through to the roof (Fig. 3.6). The computer then has to search the model. Every time it finds a component located within this given band of levels, it retrieves the appropriate 2–D image. It then displays this image at the correct orientation and plan location. In our example, it finds seven instances of the window, and Fig. 3.7 would be the displayed result.

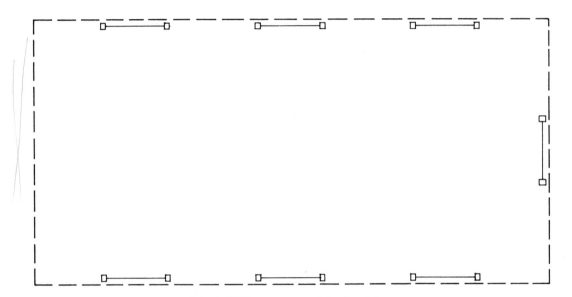

Fig 3.7 Plan generated for level 2.

Clearly we could add in to this model all the other components and building elements of our design solution – doors, ducts, partitions, cladding panels and so on. A suitable 2–D image would have to be drawn once only by the user for each component type. Then each occurrence or instance of each component could be placed in plan position, orientation and level. The whole would form a single model from which more useful layout plans could be displayed or plotted relating to any level of the building.

As the quantity of design information within our model expands, clearly it becomes important to be able to control which types of components are to appear on displays or plots. This must be done by classifying components in some manner. Then either a specified combination of classes can be displayed, or else certain classes can be eliminated from displays.

Elevations and section views of the model

We have seen that the plan views at any level within the building are produced by an accumulation process. It is done by displaying an image for a component (e.g. Fig. 3.5) at its respective plan-position. This is repeated for every component in the design solution. This works providing we have pre-defined an appropriate plan outline, symbol or other image for each type of component that has been used in the building.

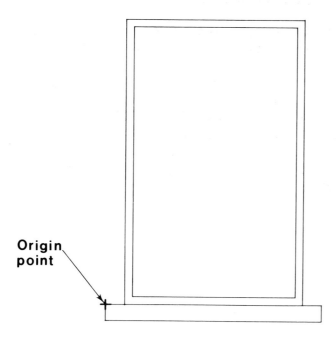

Fig 3.8 Elevation image of window.

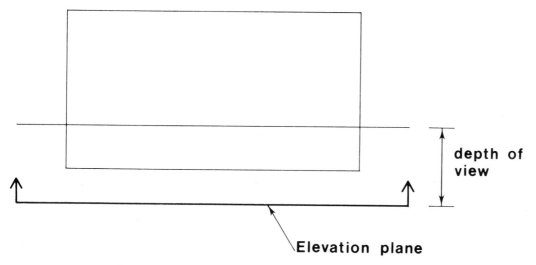

Fig 3.9 Specified elevation plane and depth of view.

So much for the plans of buildings. Let us now assume that we had instead defined an elevational image of the window as shown in Fig. 3.8. The locational model remains as before. Remember that it contains the location in 3–D space of every window in the building. Each instance however now refers back not to the plan image,(Fig. 3.5) but to the single elevational image (Fig. 3.8).

Now we could specify an elevational or section plane in the building model, plus a depth of view (Fig. 3.9). The computer would then search for every component in the model which lies within that distance of the given plane. Each time one is detected, the component image is called up, and displayed or plotted at the relevant location (Fig. 3.10).

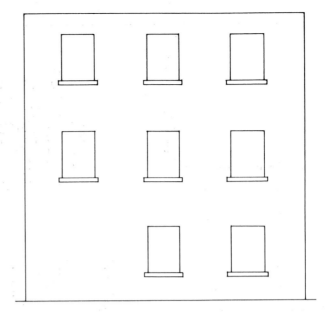

Fig 3.10 Elevation generated from model.

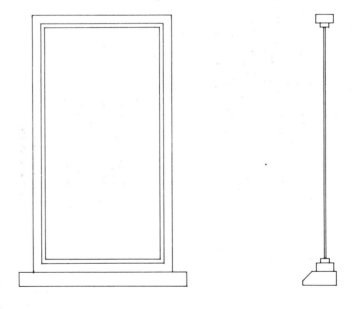

Fig 3.11 Representation of a single component by three symbols.

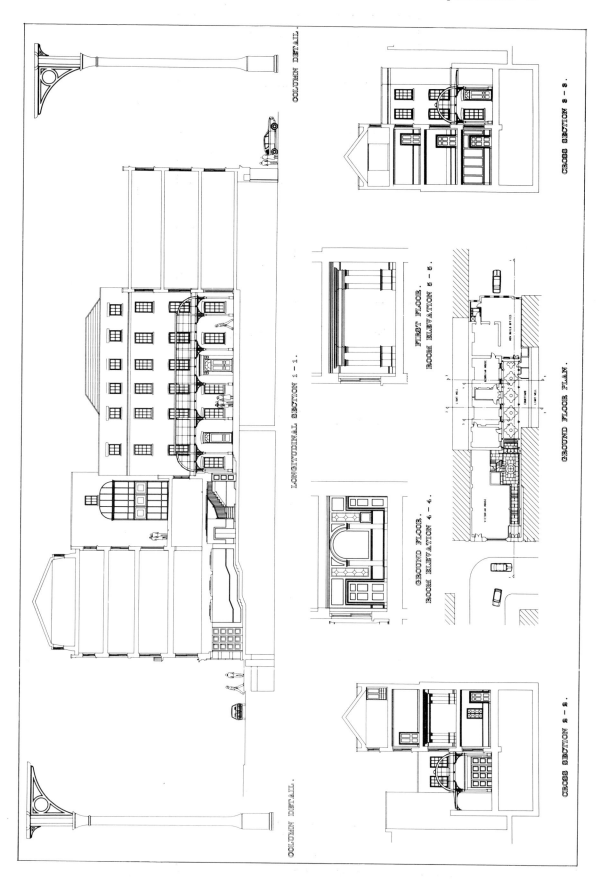

Fig 3.12 The Orangery, 161–162 New Bond Street, London. (*Courtesy*: Aukett Associates PLC, London.)

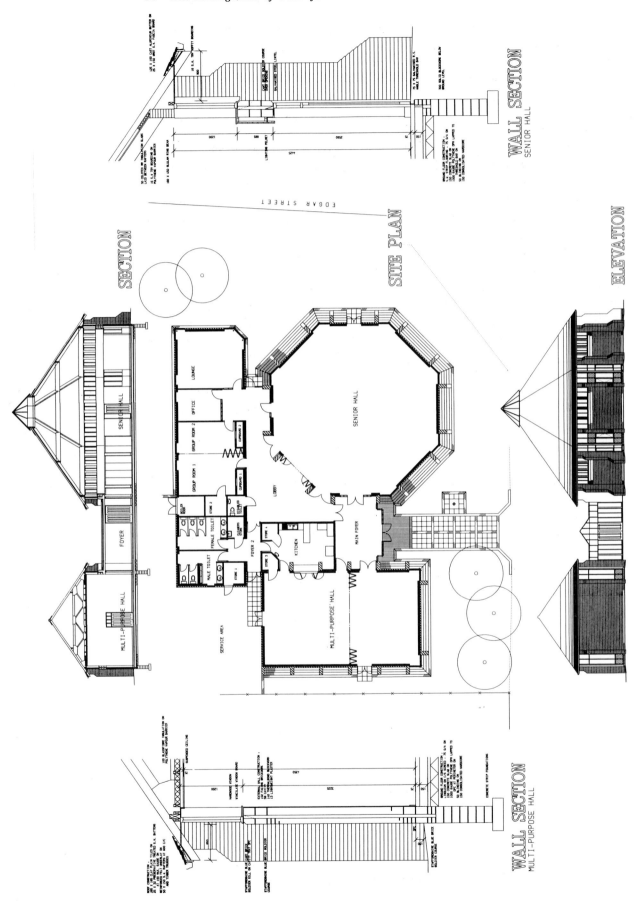

Fig 3.13 Salvation Army Halls, Hereford. (*Courtesy:* ATP Group Partnership, London in Association with the Chief Architect to the Salvation Army Trustee Co.)

A more sophisticated modelling system would allow several images, rather than just one, of each component to be predefined. They would be stored simultaneously within the computer. Thus, for example, three images might be defined and held for the window (Fig. 3.11). Then when the locational model has been assembled, the computer could generate plans at any level, elevations, or sections at any plane cut through the building. For each of such views, the computer would call up and use the most appropriate of the images which it holds for each component.

Note that each of these component images is merely a 2–D diagram which can be input using normal computer drafting techniques. The locational model can be assembled by placing components on project plans, and then keying in or otherwise indicating the level at which the component occurs. The plans, elevations and sections extracted from the single model do have the important advantage that they are all likely to be consistent with one another.

Besides representing a component with one or several images, it is possible to add component descriptions and attributes such as colour, material or cost. In this way we can build up 'intelligence' within the single locational model. With the design data co-ordinated in this way, the computer can generate several drawings and related schedules for a project with the merit of each being consistent with one another. Remember that with manual drawing work or computer drafting, each person reading the drawings has to interpret some meaning from all the lines and characters.

Note that the basic Sonata and Rucaps systems from t^2 Solutions Ltd are examples of CAD systems which generate 3–D locational model. One or more 2–D images can be used to represent each component. Figs 3.12 and 3.13 are examples of drawings produced by this method. Each is composed of the plans, elevations and section views of the buildings which have been calculated by the computer from the locational models.

3.7 THREE-DIMENSIONAL WIREFRAME MODELS

Defining the edges

We have taken a look at computer drafting, at 2–D modelling, and at a 3–D locational

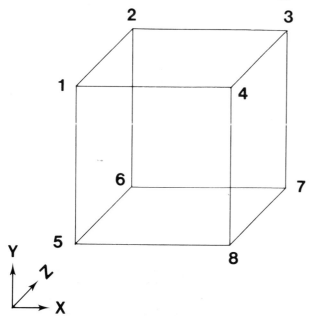

Fig 3.14 Wireframe view of a cube.

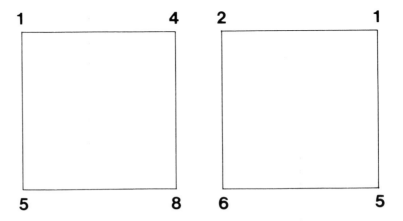

Fig 3.15 Orthographical projections of cube.

model which makes reference back to 2–D component images. Now we will look at various 3–D modelling techniques which are currently being used, starting with the simplest – the wireframe model.

In this model, each real object is represented within the computer by the lines which form its edges. Thus a simple shape like a cube would require the twelve edges for its description (Fig. 3.14). Each edge is like a wire and these are arranged like a framework.

A wireframe model could be defined by the user typing in the three co-ordinates of each node in the model, and then indicating the pairs of node numbers defining each of the edges. Alternatively the beginning of each edge-line might be defined together with the X, Y, and Z relative shifts needed to get to the other end. These methods would be laborious.

Another method would be to draw the projection of each line on two different orthogonal planes. Thus each line could be traced on the X–Y plane, and again on the Y–Z axis (Fig. 3.15). This is akin to drawing a front and a side elevation of the object. The computer will be able to determine the location in 3–D space of the line.

Viewing the model

Returning to our cube (Fig. 3.14), we must be clear that with the limited description given to the computer, it cannot recognise it as we do – as a solid cube. The computer only knows of the twelve lines in space. Programs are available which can draw views of these lines on the screen, in parallel orthogonal, perspective, isometric, axonometric or other form of projection. If for example a perspective can be created for one viewpoint, then the same program could construct a 3–D view for any other defined viewpoint. Once we have built a model, with the help of the computer and suitable software, we can look at it from as many viewpoints as we want.

Hidden line removal

Such views of a wireframe model show all the lines, even those parts that would in reality be hidden from our view. We can accept and interpret this if the view contains relatively few lines. Unfortunately models of real buildings tend to contain a huge number of lines, and the views can become congested and utterly confusing. The use of coloured lines can help in interpretation.

Programs exist which can remove all the hidden lines (Fig. 3.16). To do this, the computer must either know or make assumptions about where opaque surfaces occur

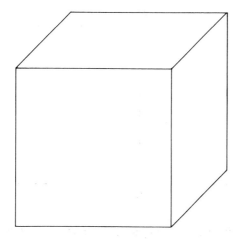

Fig 3.16 Wireframe view of cube after hidden line removal.

within the model. For example it might assume that if any closed polygon can be formed by a succession of wires, then there is an opaque surface here. On this basis the computer can calculate whether the whole or any part of each wire would be rendered invisible by any opaque surface lying between the wire and the viewer.

When the computer is left to make assumptions regarding the location of surfaces in this way, we are really asking it to extract more information from the model than exists within it. As a result the process of hidden-line removal in wireframe models tends to be error-prone. The computer in effect might assume that an opaque surface exists where there is none in reality.

3.8 SURFACE MODELLING

Defining the surfaces

The computer can be told explicitly where each real surface lies by the user indicating the bounding edges in a wireframe model. For each surface, details can also be given of its attributes such as its colour, transparency, material or cost. The result is a surface model and we have moved towards a fuller representation of the real object.

Surfaces may be defined by 'sweeping' specified lines through space. One straight line can be 'swept' linearly along another to form a parallellogram (Fig. 3.17(a)). A curved

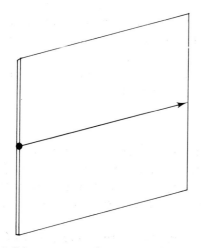

Fig 3.17(a) Line swept to form a parallellogram surface.

Fig 3.17(b) Line swept to form a curved shell.

line can form a non-planar shell surface (Fig. 3.17(b)). A rectangle can be swept to form the surface of a hollow prism (Fig. 3.18); a circle to form a hollow tube (Fig. 3.19). The sweeping action need not be along a straight path, thus sweeping a circle along an arc forms a bent pipe (Fig. 3.20).

Yet another way of defining surface models is to assemble them using certain 'elementary' surfaces. For example we might use walls, cylinders, cones, spheres for creating 3–D surface representations of real building components.

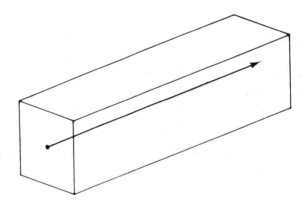

Fig 3.18 Rectangle swept to form a hollow prism.

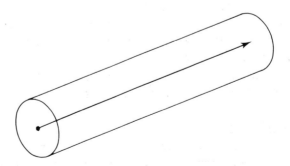

Fig 3.19 Circle swept to form a hollow pipe.

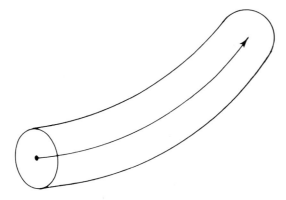

Fig 3.20 Circle swept to form a bent pipe.

Ground modelling systems are used in civil engineering, mining, quarrying and landscape design. The surfaces to be modelled here are the original ground surface, the design surface relating to the finished project and perhaps the interfaces between geological layers. The surfaces are represented either by spot heights on a grid pattern, or by 3–D string-lines which follow ground features like streams, fence lines, road kerbs or centrelines. Area and volume calculations are always needed, as are drawings such as plans, ground-sections and 3–D views. Plate 10 was produced during the design of a new reservoir scheme using the Moss ground modelling system.

Realistic visualisation

Many designers, especially architects, are interested in realistic visualisation of their models. We can now appreciate that surface modelling techniques are needed to make this possible. Increasing degrees of realism have been achieved, but only with greater efforts on the part of the software programmers, from the designers themselves at model-building time, and from the computer processor at run time. Some of the visual effects that can sometimes be calculated and displayed include:

- Colour intensity on surfaces, depending both on the level of ambient light and on the angle of incidence from one or more light sources. These are called Shaded Views. Plate 3 shows an example.
- Shadows, of an object on the ground, or else of one object on another. (See Plate 8.)
- Transparency or translucency of surfaces.
- Spectral effects, e.g. the sparkling white dot on a wine glass.
- Surface texture, such as brick and weatherstaining.
- Atmospheric effects, e.g. haze, blueing of distance.
- Photomontage – the integration of model views with photographs.
- Animation, e.g. many views calculated from successive viewpoints, converted to form the individual frames of a video or film.
- Real-time Computer Simulation. Here an adequately powerful computer system can calculate and display the changing views as the screen is watched.

Always there is a trade-off

It must be emphasised at this point that individual designers have to find a trade-off between the degree of visual realism which they feel that they require, and the cost and effort of achieving the results. As time passes and techniques improve, what can be achieved in a given time is expanding rapidly. Several examples of views obtained from surface models are included on the colour plates.

3.9 SOLID MODELLING

Solid modelling is often confused with surface modelling. Indeed many of the techniques are common and the division is blurred. Surface modellers concentrate on the definition of several surfaces. Some surface modellers are designed to cope with complex curved surface geometry. However, within the surface model, the connections between all the surfaces which form the boundary of each object might not be defined and recorded. As a result, the computer might be able to display all the surfaces and their juxtaposition so that we can interpret the objects. It might even be able to calculate individual surface areas. But it may contain no 'intelligence' regarding solid objects, or whether one object clashes with another. It may be unable to calculate volumes, mass properties, or to generate section views through objects.

Solid modelling systems essentially concentrate more on the makeup of volumes and solid objects. Solids are sometimes defined in terms of their boundary surfaces, but the connections between the surfaces are recorded. Alternatively solids may be defined by the assembly of solid building blocks like solid cubes, prisms, spheres, cylinders and cones.

The computer can calculate volumes and weights of solid objects. Likewise it can calculate 3–D views of objects from any viewpoint. It can calculate section views cut through objects, and can determine if clashes occur between objects.

Because the model is even more complete compared with the surface modeller, the assembly process is often longwinded and tedious for the system user. Also it requires much computer processing power if the solid models grow to any complexity. Usually, practical considerations mean that solid modellers have to sacrifice the ability to cope with complex-shaped surfaces.

Many surface and solid modelling systems, particularly those favoured by mechanical engineers, are reasonably competent in handling very complicated objects. A turbine blade for an aircraft engine or a helicopter rotor blade are examples.

Unfortunately such computer systems are less suitable for handling designs which represent assemblies of huge numbers of objects. The problem lies partly in the difficulty and time-consuming nature of the input operations. Also the computer has difficulty in coping with the large volume of data and with the volume of elaborate processing that is involved. As a result, solid modellers are not generally used to create detailed models of a complete aircraft engine or of a complete building – with each and every one of their components fully represented.

3.10 LOCATIONAL MODEL WITH 3–D COMPONENT REPRESENTATIONS

I described the locational model in Section 3.6. There a single component type was represented by one or more 2–D images. Each time this type of component is to be assembled into the design solution, we talk of assembling another *instance* of the component. The assembly process involves positioning the instance at its required location in 3–D space. Essentially the computer holds the orientation and 3–D location of each instance in the locational model.

A variation of this idea is to use a similar location model but to define each component not as one or more 2–D images, but in 3–D terms. Indeed, each component might be defined using either wireframe, surface or solid modelling techniques.

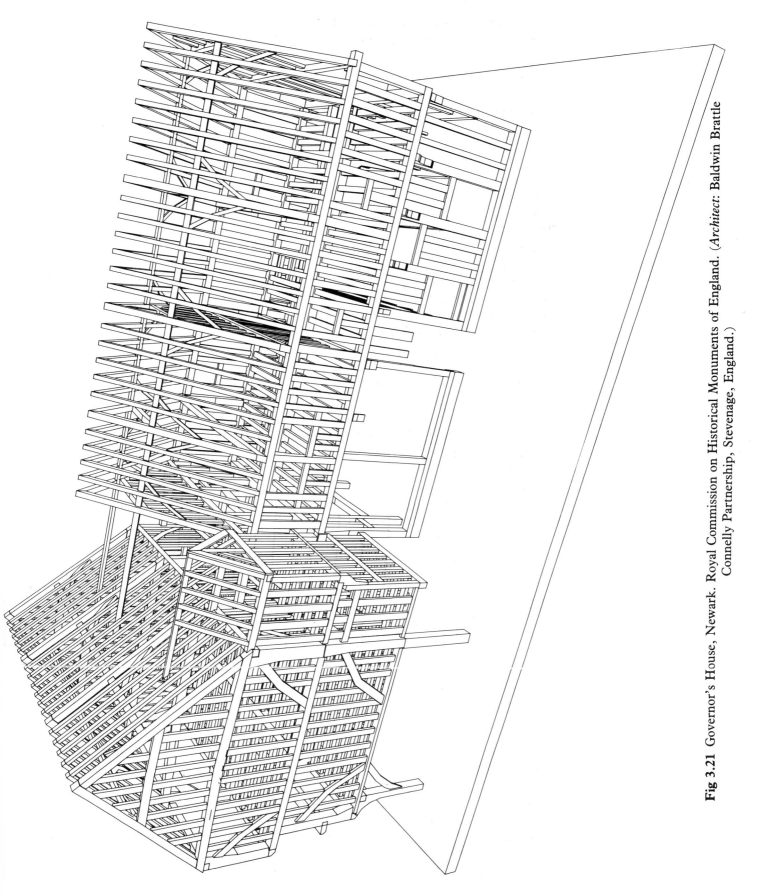

Fig 3.21 Governor's House, Newark. Royal Commission on Historical Monuments of England. (*Architect:* Baldwin Bratle Connelly Partnership, Stevenage, England.)

This form of locational model then contains sufficient information to enable 3–D views to be generated, besides the more familiar plans, sections and elevations. All such views would be consistent with one another because of course they are derived from the one source model.

The best conditions for using this technique are when the model is an assembly of:

- A moderate number components.
- Each component has simple geometry, or at least can be represented in the model with simple geometry.
- Each component type tends to recur several times.

Then the input load on the user is restricted, and the computer has less information to store and process. Figs 3.21, 3.22 and 3.23 are views which demonstrate what can be achieved.

These conditions often apply for conceptual design as we shall see in Chapter 7. They apply particularly in the building industry where projects are assembled at least partly using standard building products. The latter are not normally manufactured from information contained in the building model. Rather the model only needs sufficient detail to specify which product is to be used, to indicate its location and to provide a rough but sufficient idea of its appearance.

Representing components or elements with 'simple' geometry is often feasible in construction industry design. For example a reinforced concrete column has simple external geometry. The internal details – the steel bars – are detailed and documented elsewhere and so these need not be included in the 3–D model.

Another example is a pipe. This might be held as a cylinder with only its nominal diameter and length defined. The flanges of the pipe and joint gaskets need not be modelled because the pipe itself is a bought-in product which is not itself being designed. The joints are detailed elsewhere. It is true that an allowance might have to be made when positioning each pipe in the model for the physical size of its flanges. Nevertheless, the designer's essential requirement is to define which type of pipe is needed, its exact location, and its relationship with other components.

We shall see that this locational model containing 3–D representations of components does have some applications. These are particularly for conceptual design, and for solving certain specific detail problems. Its strength is in the ability to place and manipulate pre-defined components to form assemblies. Design errors are more likely to show up at a stage when corrections can be easily made. Nevertheless we should be clear that it is certainly not a practical proposition to apply this technique for the detail design of entire building projects. There are too many components involved and 3–D views of everything is a luxury which is neither necessary nor always desirable.

3.11 LOCATIONAL MODEL WITH 2–D AND 3–D COMPONENT REPRESENTATIONS

The locational model defines the location and disposition of each component instance within any assembly such as a building. To recap: we saw in Section 3.6 how each component can be represented by one or several 2–D images and in Section 3.10 we dealt with 3–D component descriptions.

Now let us extend this concept further by considering a locational model where a component can be defined with one or more 2–D images *and/or* one or more 3–D descriptions. The Rucaps system can provide such 2–D/3–D options.

When any plot or screen display is to be generated, the computer must understand which of the available definitions to adopt. Clearly it would use a 2–D plan image when

Fig 3.22 Governor's House, Newark. Royal Commission on Historical Monuments of England. (*Architect*: Baldwin Brattle Connelly Partnership, Stevenage, England.)

Fig 3.23 The Waterfall. (*Courtesy*: Stuart Brake. Meldrum Burrows & Partners, Melbourne, Australia.)

generating a floor plan, and 3–D geometry for a perspective. So a component like a hot-water radiator which is not required to appear on any external 3–D views, would not need a 3–D definition.

The designer can also control the selection of representation by his use of component classification and design-layer strategies. Indeed it is within his power to define a component using two or more levels of detail so that the same item can appear at different scales.

This 2–D/3–D option provides much needed flexibility. Most components can be defined in a simple manner by 2–D graphics. Schematics or symbology may be used where appropriate. At the same time, the full power of 3–D working is available and can be invoked for certain components such as those lying in the external envelope, where they are visually important.

3.12 A MODELLING APPROACH TO SUIT THE APPLICATION

It may be appreciated now that many different CAD approaches have been taken. In practice there are variations available on the themes which I have described.

Designers of turbine blades, aircraft wing surfaces, car bodies or even of fancy bottles, are interested in precise definition of complex curved surfaces. These might be mathematically defined or they might be free-form surfaces. Accurate surface definition is necessary because these designs have to be machined, moulded or otherwise manufactured from geometrical information in the model – perhaps with the help of computer numerically-controlled machine tools. The surface and solid modellers are therefore gaining in popularity among mechanical engineers working in the manufacturing area.

Table 3.1 Types of CAD System

	CAD systems		
A	2–D drafting	Uses	Diagrams: e.g. circuit diagrams using symbology. Construction details.
		Adv	Easy to implement and simple to apply.
		Disadv	Limited potential. Assists documentation rather than design.
B	2–D modelling	Uses	Projects mostly on one level, e.g some civil engineering projects.
		Adv	Simple to apply. Useful for complex one-level projects.
		Disadv	Several sub-models needed for multi-storey buildings or when elevations/sections needed.
C	3–D locational model containing 2–D component images	Uses	Detail design of multi-level/multi-storey projects, especially if component based.
		Adv	Single model, Consistent plans, elevations, and sections can be generated. Fairly simple to apply.
		Disadv	Difficulties with highly non-orthogonal arrangements. No 3–D views.
D	wireframe models	Uses	3–D representation of objects and layout of simple assemblies.

Table 3.1 (*contd*)

	CAD systems		
		Adv	The simplest form of 3–D CAD to apply.
		Disadv	3–D wireframe views are generally very complex and of limited use. They are not attractive. Hidden line removal helps but can be error-prone due to lack of information within the model.
E	surface models	Uses	For 3–D visualisations. For design of complex surfaces. Highway design.
		Adv	Realistic visualisations, sometimes high-quality and attractive, e.g. colour shaded views – internal and external – of buildings.
		Disadv	Expensive and complex to use. Only practical with rather simple models (i.e. not too many components or details).
F	solid models	Uses	Detail design of single components or assemblies involving relatively few components.
		Adv	The model with the fullest description of the components. Allows orthogonal and 3–D views, sections, volume calculations, clash detection, etc.
		Disadv	Rather complex, and difficult and expensive to use. Not practical for assemblies with large numbers of components and much detail. Not for detail design of complete buildings.
G	3–D locational model with 3–D component representations	Uses	Layout of assemblies with a moderate number of predefined components, e.g. preliminary design of buildings. Detail design of localised areas or of particular design problems.
		Adv	A practical method for the uses stated. Relatively easy to apply, especially where components tend to recur often. Realistic visualisations can be produced.
		Disadv	Not practical for detail design of large assemblies, e.g. complete buildings with all components.
H	3–D locational model with 2-D and/or 3–D component representations	Uses	Design of assemblies including buildings.
		Adv	Flexible. Caters for pragmatic choice between symbolic and schematic representation, and full 3–D geometrical descriptions where necessary. Can cater with different representational detail for different scales.
		Disadv	Multiple descriptions of some components are necessary.

The highway designer has quite different needs. He has to model irregular ground surfaces and to cope with the special geometry of road surfaces. He is interested in calculating volumes lying between the original ground surface and his designed surface. Also he is interested in calculating the lines along which these surfaces meet.

It is important to appreciate that the building designer has different problems. The surface and solid modelling techniques adopted in the manufacturing area are unsuitable for anything other than perhaps the conceptual or preliminary design of buildings. This is an area that we shall look at in Chapter 7. For detail design on the other hand, the surface and solid modelling techniques are overkill.

In practice we will find that no single modelling method can meet all needs. So there is no point in arguing about whether a particular 2–D system or 3–D system is best, when in reality variants of both techniques are needed for different purposes. As yet there is no ideal solution, but designers today can benefit from a variety of design tools in their armoury. These must include appropriate forms of 2–D *and* 3–D methods.

I shall end this chapter on computer modelling by including Table 3.1 which lists the main techniques that have been discussed. Included in this are brief comments on the main uses, advantages and disadvantages of each approach. Choice of the best modelling approach for any application in hand is an important subject to which I shall return in Chapter 10.

CHAPTER 4
Planning a CAD System

This chapter deals with the steps that an organisation ought to take to see whether or not it should obtain a CAD system. In effect, a feasibility study is needed and the composition of the small study team required to carry this out is discussed. A CAD strategy should form a part of a larger business plan for the organisation.

First there should be a preliminary study and one task is the compilation of a list of the organisation's requirements. A look at the options is relevant although most requirements are met by purchasing or leasing a ready-to-use CAD package.

Assuming the findings from this preliminary study are approved by upper management, the work can then move on to the detailed phase. This is when the products of a shortlist of suppliers are examined carefully, and the costs and benefits of implementation are studied. The system that most closely accords with the requirements must be selected. Software is the vital element for it is this that makes the computer workstations perform. However, a suitable hardware configuration is required, and an eye must be kept to see how this might be upgraded to keep in step with future needs. The ability of the supplier to support the system is also vital.

There are usually people within a company who have a deep-seated fear of new technology. There may too be some fears over job security among staff. These important subjects will be discussed towards the end of this chapter.

4.1 INTRODUCTION

A successful CAD system must be carefully planned. It does not just happen of its own accord. This chapter discusses how to set up and then undertake a CAD study. It updates material which I have presented previously[4, 5, 6].

Those designers who already have access to a suitable CAD system perhaps might skip this chapter. However, CAD planning ought be be reviewed regularly as part of an overall business strategy. Fig. 4.1 shows this as a cyclical process. Companies that are already operating a CAD system may feel in need of a better CAD management structure or better working procedures. Perhaps the existing system may have too little capacity or have limited potential. An upgraded or changed system may be necessary. Perhaps some new applications are being considered, like moving from a computer drafting system to a modelling system. This chapter should help in all such situations.

4.2 AIMS OF FEASIBILITY STUDY AND NEED FOR A PROCEDURE

The aim of a feasibility study is to find out how CAD could be applied within a firm to help designers to do their work more effectively. A realistic view of the benefits and costs

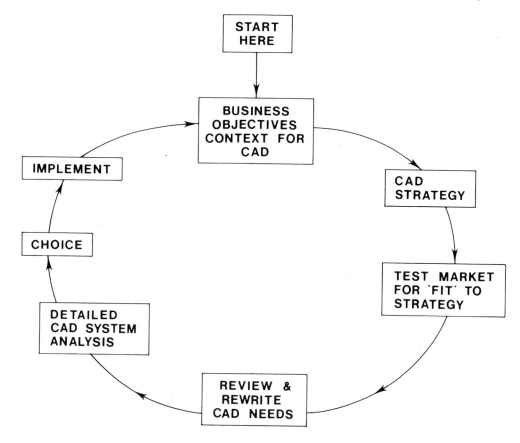

Fig 4.1 The CAD study is part of an on-going business plan.

of implementing a system must be obtained. Only then can management make a rational decision on whether the introduction of new methods would be worthwhile.

Any CAD system introduced will have to coexist with the traditional design process for many years to come. The reason is that few organisations can, or should, make a revolutionary changeover to computer methods overnight. Anyway some design work is not amenable to computer methods at present.

In seeking new facilities, it should not simply be a matter of adopting the cheapest possible CAD system and hoping that this will be the best value. Few people follow this principle when choosing their new motor-car. A CAD strategy must reach deeper than just cost benefits and superficial efficiency. CAD will touch on all aspects of the management and operation of the business, and so any study should be undertaken in a suitably broad context. Certainly the most visible of CAD costs are the capital costs of the computer hardware and software. Other costs can in the long run be more important. These are the implementation costs including training, and the on-going costs of system management, operation and maintenance. Skimping on the initial capital costs often leads to higher operational costs as well as to lower benefits.

It would be unwise to underestimate either the importance of a feasibility study or the difficulties that will be faced by those who undertake it. Many computer programs for administrative or analytical tasks affect only a few users or specialists within a firm. Such programs might be low budget items. By contrast, CAD can become the essence of design office procedure. It is likely to have profound effects that will be felt throughout the whole design organisation, and perhaps beyond. The introduction of CAD will change the working lives of people.

Industrial leaders are well used to making investment decisions relating to plant and machinery. To them it is almost a routine matter.

In a design organisation however, a CAD system may be one of the costliest investments that it has ever had to make. Decision-making is not easy. Designers are not likely to be CAD experts, while the technology in this area is moving very rapidly indeed. There are many CAD systems available and many options to choose among. All this means that sometimes it is difficult for designers to know even where to begin. There are plenty of opportunities for making serious and costly blunders. In situations of much uncertainty, there is likely to be procrastination.

Yet we should not despair for the problem is soluble. The answer is to adopt a procedure for the feasibility study. Careful planning saves both time and effort. It avoids duplication of effort, and makes a successful implementation far more likely.

4.3 THE STUDY TEAM AND ITS COMPOSITION

The start

Somebody has to initiate the process. It might be a young and enthusiastic designer who sees the need for change in his working environment. He latches on to the idea of CAD as a means of bringing it about. Let us call him our 'CAD enthusiast'.

Initially at least, he would do well to work up a few ideas on his own. As soon as possible however, he should make a case to his superiors for a proper CAD feasibility study to be started. This is because the acquisition of CAD is important enough to warrant the direct attention of upper management of the firm – the directors or partners. Investment in CAD should be a matter for the whole organisation to consider. It must not merely be a departmental matter, and for this reason it is a very unusual type of acquisition.

It is dangerous for a single person to be given the job of doing the actual feasibility study. A breadth of experience is needed and it is a job for a team of people. However progress is unlikely to be swift if a huge committee is formed, so in most firms, around three or four people should be adequate. In most cases the members will undertake these duties on a part-time basis, for their normal day-to-day work must continue without too much disruption.

The success of the whole venture depends on this team. So one of the most important jobs for the firm's management is to set up the team, and in doing so to choose very carefully from among the talent available.

In general terms, it is better to choose people who are reasonably enthusiastic and well-motivated. They will have to work to a goal that is in the interests of the organisation as a whole, rather than towards sectional or departmental interests. People are needed who can work well together within the team, as well as with others. They will have to do a lot of listening and talking, and so tact is something else which would prove useful. All such human qualities are as important as the technical abilities of the people concerned.

With regard to experience and backgrounds, the team should be broadly-based, and together cover the main design disciplines of the firm. Having looked at the ideal qualities, we can look at the composition of the team under the following categories:

- Member of the firm's upper management.
- Representative(s) from the design office.
- Representative of computer department (if applicable).
- Outside consultant.

Member of firm's management

Upper management of the organisation must be fully committed to this feasibility study, and indeed it must be seen to be committed to it. Indeed the selection and implementation of a system must take into account both the short-term and long-term requirements of the organisation. For these reasons, upper management ought to be directly represented, and so a director or partner as appropriate should be appointed to head the team. Later on, management will have to take the important decisions. This direct channel between upper management and the team should ensure a better understanding of the important issues.

Of course management people are always very busy and initial reactions are that they cannot spare the time. The tendency can be to delegate the responsibility. Where these attitudes apply, management really must ask itself whether its function is to plan the future of the organisation, or just to let it blunder on. If management cannot be seen to take an issue like CAD seriously, how can the team or the rest of the firm be expected to do so?

The management person should act as chairman, and provide the leadership and drive needed. The feasibility study is itself a project, and it too needs to be managed. Resources have to be freed to enable the study to proceed. First there is the time required by the other team members. Then the team will need access to various senior people elsewhere in the organisation for information and views. The presence of upper management within the team makes it far more likely that all these third-parties will be willing to co-operate.

Representative(s) from the design office

Representation is needed from designers, and especially from the engineers or architects who are project managers. These are key people and their experience must be the foundation on which the CAD feasibility study itself is built. They can be expected to take a very practical approach when the firm's requirements are being defined. If these people are not fully represented, then later on it will be difficult to avoid giving the impression that a system is being thrust on them. In this event, disaster can only lie beyond.

Computing expertise

If the organisation already has a computer system of some kind, or a data processing or computer department, then somebody concerned with these should be co-opted to the study team. It is possible that this person will lack experience in design work and perhaps even of computer graphics applications. This is not serious. What matters is that there may be opportunities for integrating computer hardware functions, and technical questions may arise concerning data communications and computer procedures within the firm.

On no account however, should overall control of the system selection and implementation be wrested from the design professionals, to be given to data processing or computer people. Computer expertise is required to support the design professionals, not to rule over them. No empire building on their part should be tolerated.

Clearly, if there is already a computer drafting or design system in operation – perhaps used by other design disciplines – then again the opportunity for integration and of building upon kindred experience should not be missed.

If the firm lacks much experience with computers, then it certainly should not be deterred from becoming involved with CAD. The company can make progress even if it has little computer knowledge, provided members are open-minded and willing to learn.

CAD consultant

In advocating the inclusion of an outside consultant in the team, clearly I have to admit to being biased. It is certainly possible to find and install a CAD system without such assistance, and some firms do proceed in this way. However, the chances of success and the speed with which it can be achieved are enhanced if proper and relevant outside experience is brought to bear. It is in the area of introducing new ideas and new technology to the firm that management should be especially ready to look for external help.

Small organisations are particularly in need here. But large firms frequently employ consultants even when they seem to have all the relevant expertise. In the large companies the scale of the investment is likely to be bigger. Also the right consultant can help to breathe fresh ideas into the organisation and broaden the total experience span of the team. He can act as a counter to powerful internal politics. This assumes that the consultant is impartial in his attitudes, has a track record, and has up-to-date CAD knowledge. The CADCAM Association[7] identified a number of guidelines for picking a consultant, under the headings:

> independence
> experience
> ability to understand your problems
> ability to communicate at all levels
> ability to translate the academic into reality
> academic qualifications
> don't be afraid of age
> price

There should be no confusion between a computer expert and a CAD expert. The latter combines knowledge of computer graphics techniques – rather than just conventional data processing – with experience of design procedure.

The consultant should not be allowed to do all the work independently or take any major decisions. Decision making is the province of management alone.

4.4 OUTLINE OF PROCEDURE FOR FEASIBILITY STUDY

I suggest that work of the team should be divided into three stages:

1   Preliminary study

_____ Preliminary Decision Point _____

2   Detailed study

_____ Main Decision Point _____

3   Place an order for system

The decision points mark the times when a report has to be made back to upper management. It has then to make decisions, with the benefit of the team's advice, on whether to carry on to the next stage. Essentially management has to weigh up the likely benefits of proceeding, and compare these with the further expenditure that it would have to sanction – on a detailed study or on purchase of the chosen system. In discussing the feasibility study here, I am assuming that these decisions are positive.

The whole feasibility study is rather like a commission to design a building project, where the equivalent stages would be the preliminary design, detailed design and tender procedure. Like the design process, the study is a very iterative procedure. There is always an intermixing of activities, with the work gradually becoming more and more detailed. I am able to separate out the activities here for the purpose of explaining them, but such separation is not so easy in practice.

Progress depends very much on the individuals in the team. As a result, management must be ruthless about replacing any members who in the course of the work do not prove to be suitable.

4.5 PRELIMINARY STUDY

The main tasks are:

1. Raise level of awareness in CAD.
2. Self-analysis to identify the CAD applications within the organisation.
3. List the CAD objectives and the requirements of the organisation (in writing).
4. Study the options.
5. Make a preliminary assessment of the benefits deriving from the introduction of a system.
6. Make a preliminary estimate of the costs and compare these with the benefits.

Awareness and applications
No opportunity should be lost for gathering more information on CAD techniques. I hope that other chapters in this book will be useful, while other potential sources are exhibitions, conferences, technical meetings, books, magazines, and meetings with designers working in other companies. After some information has been absorbed, enquiries should commence into the organisation's structure and workload. Gradually the team has to form a judgement on the type of projects and work areas where a CAD system could be introduced. It is obviously best to concentrate most on those work areas which are central to the firm's activities, and avoid peripheral interests.

Objectives and requirements
Objectives should be reviewed at two levels:

1. Objectives of the organisation.
2. CAD objectives and requirements.

The second of these is perhaps the most obvious task for this study team. However, it can only deal with CAD objectives if it knows where the whole organisation is going. The chairman of the study team may be able to throw some light on this, but unfortunately the management of too many design firms have little notion of this.

With regard to CAD objectives and requirements, the team should start to note down some ideas on paper. These should *not* be just in terms of the end-products, like hardware devices, that will be required. Too often, people identify their needs as say, a computer and A–0 plotter. They may well need these as incidentals, but such items in themselves will solve nothing.

The team ought to start taking a stance on whether it would be happy with a 2–D draughting system, some kind of a modelling system, or if some combination of techniques would be more appropriate (refer Chapter 3). The size and complexity of the projects tackled, and the quantity and nature of documents which are to be the final output should be studied.

Initially it is not easy to note the objectives and needs on paper, but it is certainly worthwhile to try. The attempt itself forces the team to think deeply. Its job is to identify the real problems to be solved, rather than imagined ones. It focusses attention and makes it more likely that the team will identify a system which will help the firm rather than one which becomes a millstone or a financial liability. Initially the team's list may be very tentative and short. But the team should expect to keep on changing and improving it as the studies progress. Indeed the list may never quite stabilise. Later on it will become the yardstick with which particular solutions are compared. Individual requirements noted in the list will have different values or weights, some items being rated as essential, others as merely desirable.

Options

Since organisations and applications differ, the team will create a unique list of objectives and needs.

The most likely solution is for the organisation to acquire a 'ready-to-use' CAD package. There are many suppliers. Each tends to offer one principle software module, plus other optional add-on modules suited for particular applications.

With its software, a supplier can normally offer a range of hardware options, although these might be restricted to the models produced by just one or two hardware manufacturers.

By selecting from the hardware and software options a supplier can offer a system which is tailored to some extent to meet a customer's individual needs. Thus a supplier of such packaged CAD systems can become a single source to a customer for:

- all the hardware and software,
- operator training,
- on-going support and maintenance of the hardware and software.

The study team might well believe that it would be prudent to phase in a system over a period of a few years. This eases the cost burden. Although it might impose some delay in the reaping of the ultimate benefits, this reduces the risk of too much disruption to the on-going work of the organisation. Adoption of a phased approach makes it essential to pay attention to the ways in which it is possible to upgrade the system in a step-by-step manner.

Occasionally a firm may feel that it can do better, and perhaps save some money, by configuring its own system. This means that it acquires individual elements of hardware and software from different suppliers and itself puts them together. This option is only advisable if the firm has access to plenty of computer expertise. Even then, any initial cost savings might well be offset by difficulties experienced in linking the elements, and in supporting, maintaining and developing the resulting system in the longer term.

A few firms might have such specialist requirements that the normal ready-to-use packages are either not applicable or else they would be too inefficient in application. For example they may have routine applications which link graphics or drawing with special analytical procedures or materials procurement processes. In such circumstances it is possible to commission special software to be developed. This would initially be a very expensive option, and would only be justified when the potential benefits are even larger. Big organisations having their own software development teams might consider the development of graphics software in-house, but again only if the available ready-to-use packages are quite unsuitable. In these circumstances, the development team would do well to acquire a graphics subroutine package first[8] and base the development of their application software on this. Then the in-house software development team could concentrate most efforts on the technical application to be solved, rather than dissipating valuable resources on the programming of the graphics effects.

An alternative to acquiring a CAD system is to use the services of a CAD bureau. This might well prove to be an expensive option if it was adopted over an extended period of several years. However by using a bureau for a shorter period, a firm can quickly assess a proprietary CAD system and acquire some experience of operating it without committing itself completely by a system purchase. Thus at a cost, this removes some of the risk from the CAD implementation phase.

The bureau solution is not too straightforward however. Either:

1 the customer has to provide the operators and arrange for them to be trained. This takes time, and the operation could prove costly while they are gaining initial experience; *or*
2 the operators employed by the bureau have to be used. This creates problems over communications with the designers, and perhaps these will be exaggerated by physical distance.

In some instances, initial high hopes descend until the bureau system is acting as little more than a tracing agency for the firm's drawings.

It is possible to install terminals within the firm's offices which are linked with the bureau computer. The dial-up mode of telecommunication links would not be adequate for CAD work however. A fixed telecom link is usually necessary, but this adds to costs and might need some time to be put into effect.

In the majority of cases, a look at all the options will probably result in the team recommending that a ready-to-use CAD package should be acquired. So this solution is the one which I intend to concentrate on.

Preliminary view of costs and benefits

A preliminary idea of costs can be gained by talking to one or two suppliers, and by studying CAD literature. Note that system prices quoted in magazines are often misleading. Usually they relate to the most basic system available, and of course this might not be suitable in the relevant circumstances. Also the cost of a reasonable plotter, itself an expensive item, is often omitted in published figures. The price of training courses may, or may not, be included in system prices and these too can be quite significant. The system acquisition cost also depends on how the purchase is to be funded.

The feasibility study itself might cost 10% or more of the system capital cost. Then from the time of delivery onwards there are the costs involved in:

- system management,
- operation staff,
- accommodation for equipment and staff,
- consumable materials,
- insurance,
- support and maintenance from the supplier.

The maintenance and support is covered by a separate contract, usually renewed annually. Each year the payment would be around 10% to 12% of the capital cost of the entire system.

All these costs are obviously incurred at different times. Since we need to compare costs with the benefits of operating the system, we should include only the *additional* costs incurred compared with existing practices. Also each of the costs that will be incurred in the future, and the perceived benefits too, ought to be discounted to their present-day values. The purchasing company must be prepared to face the fact that the implementation, operation and management of a system over a few years will cost more than the initial system price.

With estimated costs established, it is possible to calculate the cost in use of each workstation-hour. For this we must make an assumption regarding the utilisation of the stations. In the absence of shift working, a figure for example of five hours per working day for each available workstation might be a reasonable assumption.

With a computer drafting system that closely mirrors the manual drawing practice, the cost of operating the new system can be compared with the traditional drawing practice. This *cost ratio* of the new and old methods is:

$$\frac{\text{cost-in-use of workstation/hour} + \text{cost of operator/hour}}{\text{Cost of draftsperson/hour}}$$

The operator's and draftsperson's costs can be estimated from the relevant salaries with a factor to allow for overheads.

This cost ratio is a simple measure of the amount by which drawing productivity would have to be increased if the computer system is to be financially viable.

This initial analysis has ignored the intangible benefits of CAD, a subject to be discussed in Section 4.6. These may be far more important than productivity gains and this is why too narrow a view of costs and benefits can lead to wrong decisions. At this preliminary phase, the team should begin to evaluate qualitatively all the benefits that might accrue.

4.6 PRELIMINARY DECISION POINT

The preliminary study is necessarily rather superficial. When it has progressed a little way, and some initial views are beginning to gel, then is the time for a reappraisal both by the team and by the firm's management.

Is the team looking for the right sort of solution? If it is already clear that the benefits for the particular organisation are not adequate, or if the money available will simply not match the sort of system that the team think is required, now is the time to discover and face the fact. Perhaps it would be better to stop before too much time is expended.

More commonly, after a critical reassessment, the team can embark upon the detailed study with a renewed mandate.

4.7 DETAILED STUDY

The detailed study is really more of the same thing, but literally everything needs to be taken to a deeper level. Now the potential suppliers and their products need to be examined. The study team's list of CAD objectives and the organisation's requirements must continue to evolve throughout all this time.

When systems are demonstrated by the vendor, the buyer must try to exert some control over the proceedings. This can be done by assessing the demonstrated system against this list. The question that should never be far from the surface is 'Would this system provide what we need?'

Further information on assessment of systems is given elsewhere[4, 9]. Some large companies devise some form of 'benchmark' test to help in their evaluation of one or two short-listed systems. The difficulties involved, and skill needed to perform effective benchmark tests should not be underestimated[10].

Valuable information can be gained during visits to reference sites where the system under examination has been in regular operation for several years. If such reference sites do not exist or cannot be found, the system has no track record and a purchase must be too risky to contemplate. CAD technology is still somewhat immature, but there are already several established systems from which any customer can choose.

Most of the team's attention should be fastened on the software, for it is this that determines whether the hardware boxes will be useful or not. If the software is right, then in most cases it will be possible to work out some configuration of hardware devices to suit the individual circumstances.

Planning the hardware configuration

After the software has been assessed, it is then time to look with the potential supplier at each and every item of equipment that would be needed. This examination must get down to the hardware model and capacity so as to arrive at the precise specification and price of a complete workable configuration. For this to be possible, the customer needs to have some idea of the nature of the workload that will be committed to the system.

The 'sizing' of computer hardware is hardly an exact science. It relies heavily on judgement and advice from the supplier. It is very unsatisfactory to implement a system with inadequate capacity, but conversely it is also very wasteful to fund under-utilised hardware.

It is always worth checking this out by looking at the make-up of one or two systems that have been installed recently in other organisations by the same supplier. When doing so, bear in mind that the needs and applications of these organisations will not be identical. Also take into account the age of the system. In the initial month or two, most users are gingerly inputting simple commands and working with small design models. The computer is not being taxed heavily, and so it may seem to respond very rapidly. After many months the situation becomes radically different. Then the users are much more capable, and more impatient too, since the novelty has worn off. The design process has moved on too and they are trying to work with far larger models. The computer may now be over-loaded for much of its working time. The symptoms are reflected in slow response, operator impatience and low productivity. Systems must be configured to cope with these, rather than with the initial circumstances.

Upgrading in the future and distributed systems

A system has to be selected and configured now, but this system will certainly have to be upgraded at some time in the future. Indeed, it is usually best to start with a relatively small configuration and plan to let it grow by adding more workstations and other facilities as experience is built. However, upper management will have to clearly understand and be sympathetic to such an approach.

It is therefore imperative to take a hard look at the systems potential for growth and development:

1 to provide more capacity, *and*
2 for new applications.

By the time an upgrade is actually needed, there may be new hardware models or software items available. Completely new options might have opened up. However if plenty of upgrade options exist now, then it is an open and flexible system. With this, the design office will be better prepared for the uncertain future. Conversely if there appear to be few upgrade options now with a system under investigation, there may be still fewer options when an upgrade is actually required in the future.

With single-user systems it is of course possible simply to buy one or more new systems and to run them all in parallel. But the means by which the data can be passed among the individual systems requires some investigation.

Design offices which are of any significant size should look carefully at the merits of *distributed systems*. Multi-user CAD systems are increasingly being configured as user-workstations, each of which contains its own processing capability. Several of these can be linked within 'networks' to provide:

1   communications to other users working on the same or similar projects,
2   access to a shared plotter,
3   access to large discs for large-scale data storage,
4   access to more powerful processing capacity for occasions when this is needed, e.g. structural analyis, energy calculations, simulations.

This concept of distributed systems has already become the predominant computer configuration for CAD and other applications.

With distributed systems, the total hardware solution is able to evolve over a long period of time to keep in step with the needs of an organisation. We can add a new item here, replace an existing item there, trade in another item for a new one elsewhere, move an element physically to a new site, and so on. At last, the idea of investing in computer equipment which rapidly becomes obsolete and then has to be replaced, is itself becoming an outmoded idea. In its place, we can invest in a small system initially, and keep abreast of needs by altering it to suit requirements. We would be well advised therefore either:

1   to install a distributed system from the outset; *or*
2   to implement hardware that has the proven potential for integration into a distributed system in the future.

Also we should start to think of investment costs in terms of on-going current expenditure, rather than as a huge one-off capital investment.

4.8 COSTS AND BENEFITS

By this time, the system selection will have narrowed to just one or two options. Configurations will have been planned in some detail. The initial estimate of costs – including on-going costs discussed in Section 4.5 – can now be updated with more accuracy.

The ratio of CAD workstation operating costs (with operator) to the cost of a draftsperson doing drawing work manually (see the Cost Ratio discussed in Section 4.5) will indicate the drawing productivity factor that must be achieved to ensure that the system will be economically viable. This productivity factor is the speed with which drawings are turned out with the system compared with doing similar drawings manually.

A more elaborate approach is to make a reasonable assumption regarding the probable productivity ratio. With the value of these drawings and the cost of the system known, the internal rate of return of the proposed investment may be calculated. This method is preferable because in principle, management has to make sure that investment is directed towards the best of the many conflicting needs for the available money. Their choice must be the investment which is likely to provide the highest rate of return.

Drawing productivity and factors affecting it

Productivity trials with one particular system were carried out in 1982[11]. These embraced general arrangement drawings, structural outlines and details. The indicated productivity ratios achieved were between 1.5 and 4.5. For reinforced concrete detailing, the same firm achieved ratios ranging from 1.25 to 2.8 with an average of 2.2. The cost-benefit break-even point was judged to be 2.0, this being based on then-current costs of hardware, software and staff. So at the time it was marginally worthwhile using computer methods. Of course this is comparing costs incurred with the single benefit of improved drawing productivity. All other benefits of computer drafting were ignored. In the years since these trials, hardware and software have improved while their costs have fallen, and staff costs have escalated. As a result, CAD must be a more attractive alternative at this time.

Productivity ratios do vary considerably, perhaps even more than suggested by the above figures. Productivity is a complex issue. It depends primarily on the quality of management of the CAD resource. Productivity also depends on the nature of the system employed. All other things being equal (and they are not always equal), a powerful system with plenty of useful features will of course cost more than a basic system on a small computer. However, benefits in the form of higher productivities are likely to be achieved with the former.

With a modelling system, as opposed to a drafting system, much of the operators' effort is in component creation and model building. With this work done, relatively little extra effort is needed for creating each of the derived drawings. A productivity ratio must however be gauged over a whole design stage rather than on a drawing sheet-by-sheet basis.

Productivity is also dependent on factors like the nature of the work (e.g. the amount of reuse of components that is possible). Likewise it depends on the level of expertise of individual operators, and on their mental attitude and state of alertness at any given moment. Operator skills and attitudes are more important important in CAD operations than in manual working.

Productivity benefits will be low or negative in the early weeks following the introduction of a new system. The benefits will steadily build up over the months as the skill levels of operators improve. Benefits will also mount as more of the firm's experience in built up in a tangible form. Much of this can be held within computer-based procedures, libraries of symbols, components, details, standard notes, drawing frames and the like.

It is apparent from this discussion that the major factor which determines how productive a particular system will be is the quality and effectiveness of management. By management here I include three variants:

- Practice management – the organisation.
- Project management – the design process.
- System management – the hardware, software, data, operators, support and maintenance.

If there is little or no management, then productivity will certainly be low, and furthermore it would soon become obvious to all that the investment has been wasted.

Other benefits from computer modelling

Design quality: This can be improved as a result of design iterations carried out interactively with the computer. The prototype models of the project can be more thoroughly investigated and assessed. The project itself may be easier and cheaper to build and maintain.

Design quality and reliability are important. Companies expect quality and reliability from their suppliers, and consumers expect it in their purchases. Clients' organisations are looking for it too.

Marketing opportunity for user company: Many clients already perceive that the use of CAD on their project is in their interests, and this can affect their choice of design firm. The capture of a single extra project might pay for the CAD system.

Documentation quality: Drawings tend to be by-products from the model, and it can be easier to produce clearer and more complete drawings.

Consistency of information: Due to the single-source model, the information contained on multiple sheets of drawings is more likely to be consistent.

Design process is speeded up: Design work can be done more rapidly, and so the design cycle for projects can be shorter. The design office can respond faster. Time is money. Quicker project design also means that office cost overheads accumulate over a shorter time. Clients can start to reap the return on their capital somewhat earlier.

Dimensional precision: The CAD system can operate at far greater accuracy compared with what is possible on the drawing board. The dimensional control for a whole project is enhanced.

Opportunity to improve procedures: Implementing a CAD system provides an opportunity to institute new procedures where necessary, or to improve existing ones. Examples of such procedures are those for design checking, drawing approvals and revision control.

Design modifications: Design solutions that are subject to much change can be accommodated more easily. The work of changing the model may be far from trivial, but when this is done all the drawings that need to be revised then tumble from the system more or less as quickly as they can be plotted.

Retrieval of design information: With modelling, it is usually easier to find design information quickly. It is in the design model rather than spread over a multitude of pieces of paper, some of which might already out-of-date.

Job enrichment: For all the staff closely involved with the application of the system.

Interestingly, some of the benefits clearly accrue to the owners and users of the facility being designed. Benefits from fewer design errors accrue mostly to the builder, while higher-quality documentation benefits all those who will use it. So the types of organisations that can benefit most obviously from investment in a modelling system are:

- those large authorities that design, own and operate their own facilities,
- construction companies that provide design/build services,
- multi-discipline design practices,
- Single-discipline practices looking for a design-edge, enhanced reputation and marketing advantage.

The cost-benefit comparison

We have looked at system costs, and discussed productivity and other benefits which can arise from use of a CAD system. Unfortunately it is one thing to discuss them qualitatively and quite another to pinpoint them quantitatively in a particular situation.

When a company buys a new machine-tool for a factory, its production rate can be determined fairly accurately. Given an assumed load factor and a knowledge of the value of the output products, then the benefits of using the machine compared with existing production facilities can be calculated fairly easily. Rational investment decisions can be made in those circumstances.

With CAD, the cost-benefit calculation is no different in principle. However the rate and value of the output products are not easy to estimate. Lawrence[12] looked at the economics of established CADCAM systems and found that there is a distinct lack of evidence to quantify the real benefits. Most users were either generally happy with their investments, or believed that the failure to get the full return was to some extent their own fault – with management of the systems singled out as crucial.

Too many design offices are apt to regard the main end-product of their work as the drawings and other design documentation – rather than the design solution for the project itself. As a result, most emphasis is on turning out drawings as rapidly as possible. Hence their interest in the simplest computer systems which they believe show the most tangible benefit to them – improved drawing productivity. They opt for a drafting system instead of looking beyond for something with more possibilities. It does seem that undue emphasis on the most tangible of the benefits – on drawing productivity – is both narrow-minded and short-sighted.

Accountants' inspired demands for a quick return on investment of three years or less may or may not be fulfilled. Such demands ignore the larger question of whether or not the design office will be in a fit state in three years' time to compete for business.

The intangible nature of most of the CAD benefits does create real problems for us. In reality, normal cost-benefit methods tend to break down because some of the benefits cannot be quantified easily and entered into a cost-benefit comparison. What frequently happens is that to satisfy a narrow accounting mentality, and salve consciences generally, over-detailed figures are conjured up which would never stand up to any really critical examination.

It is better to recognise the difficulty and first of all make a reasonable estimate of the costs. This certainly can be done. Initially these estimates will be rough. The accuracy can be increased later after a system has been selected and the hardware configuration planned. Reasonably accurate costs are always needed by management because it has to set aside the financial and other resources that will be needed.

With regard to the benefits side of the equation, the team must think as deeply as possible about all the benefits, not just about drawing productivity. Some reasonable estimate in financial terms should be made of each benefit, without any false claims being made about accuracy. With more confidence, the team must now rework its earlier cost-benefit study along these lines.

4.9 MAIN DECISION POINT

By this time the team must have started to think deeply about how the proposed system will be implemented (see Chapters 5 and 6). Also it must have taken a close look at the ability and capacity of the supplier's organisation to assist in the implementation, training, maintenance and support of the system on a semi-permanent basis.

The team is now armed with a better appreciation of the economics of the proposed investment and has a deeper knowledge of CAD practice.

When the team has made up its mind about which is the best system for the organisation, the time has come for the second major decision – whether to go ahead and place an order. Their job is to advise upper management.

In presenting their case, the team must summarise its recommendations and explain the reason for making them. Their cost-benefit figures must be augmented by a carefully reasoned argument which describes all the significant, though intangible, benefits. An understanding by management of the issues is vital. Armed with this information, management ought to be prepared to make a subjective judgement of the cost/benefit comparison, and base their decision on this. After all, many investment decisions are made largely on the basis of subjective judgement. Examples are the choice between optional schemes for a new building, which motor-car or which house to buy.

4.10 ORDER THE SYSTEM

The supply and maintenance contracts
There are two contracts to negotiate and agree with the supplier.

1   For the supply, delivery, and commissioning of the hardware and software, and for the provision of initial training of users.
2   For the supply of on-going support and maintenance of hardware and software. This contract will be subject to annual renewal.

Both parties have to understand exactly what is required of them and what to expect. A little care and attention to detail now is likely to pay dividends – possibly in fewer recriminations and misunderstandings – later on. I have covered the subject of these legal contracts and the necessary financial arrangements elsewhere[4]. The customer would do well however to seek proper legal advice during the purchase stage.

4.11 FEARS OF NEW TECHNOLOGY AND JOB SECURITY

When CAD is being considered, fears of the unknown, of new technology and of job security are usually deep-seated amongst at least a proportion of the staff. Often the most affected are the senior or experienced members of staff.

Although there is often little real foundation for such fears, it is far better if the problems are recognised and indeed anticipated. It is the job of management to make the position plain, and the team carrying out the feasibility study will have to advise management on the matter. It is a matter to be taken seriously because resentment and active resistance can kill a CAD system and waste the investment.

First, we should recognise that we are all happier with the things that we are familiar with, and so are naturally resistant to change. People are inevitably committed to the old systems in the firm. In particular the more senior staff are likely to feel in command and to have control over their work situation. Designers who have acquired some experience know that this gives them an advantage, a definite edge, over the youngsters that lack such experience. The introduction of CAD can appear to threaten all this.

This senior/junior divide is potentially very serious. Often it is the juniors that are keen on new technology. They may already have used computers and seem to know all about them. As a result they are sometimes appointed as operators. They then receive the operator training while the others are left in the cold, not knowing quite what it is all about. Experienced designers fear de-skilling; that young designers armed with a CAD system will soon overtake them in ability.

Such fears are based on a misunderstanding of:

● computer knowledge,
● CAD knowledge, *and*
● design experience.

These are all different things. First let us look at computer knowledge. Modern CAD systems are now packaged in such a way that users need to have little real knowledge of how computers work. This is just as well because so often the computer expertise of some youngsters is little more than finger agility anyway. Experienced designers are busy people and should not feel too embarrassed by a lack of detailed computer knowledge.

CAD knowledge is another matter. A designer or technician can learn the mechanics of CAD operation in a week or so. The build-up of operating experience takes a few months or even a year or two – depending on what level of experience we are talking of. An awareness of how CAD can be used is rather different again to operation of the workstation. Having some design experience helps. This book is all about using CAD, and the references and bibliography are further sources of CAD knowledge.

Architectural or engineering design experience takes years to learn!

So our computer-literate youngster who seems to have all the advantages in life has in fact got to learn about CAD operation, and gain design experience. The able and experienced senior who fears being cut down to size and even being put at a disadvantage has only to do something about his lack of CAD knowledge. It is his irrational fear that is the first thing to tackle.

Job security is a subject that needs some cool assessment. If indeed productivity is to be improved and if the workload of the design office remains steady, then some jobs are going to be eliminated. Other jobs in the design office will certainly change. It is all part of the pattern of business evolution. Firms that face the difficulties and try to do something about them are likely to fare better than the ones that choose to bury their heads in the sand and retain the old methods. These last are the firms that may suddenly disappear altogether when they can no longer face up to the competition. Then everybody there loses their jobs, so there is no security in that course.

This is a pessimistic view, for it assumes an unchanging workload. If the design office can become more effective with the help of modern technology, then it is likely to pick up more work. The jobs which disappear or change due to the introduction of CAD tend to be the ones with low skill levels. So management should start thinking about how these endangered people can be retrained so that they can increase their skills, and move into more interesting and productive jobs.

Anyone who doubts that there is enough work to go round need just look around at the country's infrastructure. Much of it is old or even crumbling. Society needs all the able designers there are to help to upgrade the building stock. Above all, the quality of the finished projects has got to be improved. Increased complexity in new buildings as well as the growing interest in quality assurance are manifestations of the fact that much more effort is going into planning and design services than hitherto.

Anyone who thinks that on a world scale there is a shortage of work for architects and engineers should read the paper by the late F. Andrew Sharman[13] who in 1985 wrote:

'Sober, well informed estimates for the next 15 years produce some very startling expectations. By the year 2000, 1400 million *more* human beings will be added to the 1800 million who were found in the world's cities in 1980. Of these 1400 million, 259 million will turn up in the industrialized 'north' countries of the Brandt Commission's reports, and no less than 1144 million will be added to the 1980 total of 972 million in the developing world, or the 'south'. The world in 2000 will contain 60 cities with more than 5 million inhabitants.'

Referring to the needs to the 1400 million additional city dwellers, Sharman points out that these are:

'. . . requiring say 300 million extra dwellings, which will all need – whether they get them or not – water supplies, schools, drainage, electricity, roads, markets, mosques, churches, local government, transport systems and hope.'

SYSTEM PLANNING CHECKLIST

1   Management to produce a business plan and CAD strategy; then to initiate a CAD feasibility study.

2   Management to set up a study team.
   (a)   Member of firm's upper management.
   (b)   Representative(s) from office design disciplines.
   (c)   Representative of computer department (if relevant).
   (d)   External expertise (if required).

3   Team to carry out preliminary study.
   (a)   Raise its level of CAD knowledge and awareness.
   (b)   Identify CAD applications.
   (c)   Make a list of CAD objectives and requirements.
   (d)   Study available options.
   (e)   Make preliminary assessment of potential benefits.
   (f)   Make preliminary assessment of initial and on-going costs.

4   Team to make preliminary report to management.

5   Management decision on whether to continue feasibility study.

6   Management should initiate its thinking about someone to co-ordinate its CAD activities (see Chapter 5).

7   Team to carry out detailed study.
   (a)   Items 3(a) to 3(d) to be examined in more detail.
   (b)   Identify shortlist of suppliers.
   (c)   Attend demonstations of systems.
   (d)   Visit reference sites.
   (e)   Benchmark system(s) if necessary.
   (f)   Assess software.
   (g)   Assess corporate hardware strategy and upgrade potential of offered systems.
   (h)   Plan initial hardware configurations in detail.
   (i)   Choose the hardware/software solution which best matches the team's list of requirements.
   (j)   Rework the cost/benefit analysis.
   (k)   Study methods and procedures for the implementation and managment of the system (see Chapters 5 and 6).

8   Team to report to management with recommendations.

9   Management to decide to invest (or not) in a CAD system.

10   Management to communicate with staff and cope with fears of new technology and of job security.

11 Management to initiate measures involved in the implementation and management of the CAD system (see Chapters 5 and 6).

12 Agree on contract with system vendor for supply, delivery, and commissioning of hardware and software, and provision of necessary training.

13 Agree on contract with system vendor for supply of on-going support and maintenance of system.

CHAPTER 5

Implementing a System – Management Issues

This chapter deals with the manner in which the people need to organise themselves to ensure that the new CAD system will function effectively.

The need for management in design practices is not always self-evident but the new pressures which design firms are experiencing are described. The merits of different kinds of CAD management structure for firms are covered.

Three CAD management roles should be assigned, namely a CAD Director; a CAD Co-ordinator to deal with the application of the CAD system to project work; and a computer manager to deal with the hardware, software and data matters. The assignment of these three roles, and the responsibilities attached to the first two are defined. The responsibilities of the computer manager will be dealt with in more detail in Chapter 6.

Operators have to be selected and trained. The subject of training is explored in detail for it must be given a high priority. It must cover much more than just the initial operator training, and include for example the awareness training of project designers.

Workstations are normally shared and a pattern of work-sessions established. Sometimes the working day is extended by adopting flexible working or shift arrangements. A scheme for workstations rotas has to be instituted.

So this chapter deals with practice management issues. The next chapter concentrates on the system management – the technical issues of implementing the hardware and software. Chapter 8 covers the project management and planning using a CAD system.

5.1 INTRODUCTION

Historically, building and construction has been a slow-moving and low-technology industry. Project design was a 'known' process. Designers knew roughly what was expected of them. The procedures adopted on one project were much like those needed for the next. The few individuals involved would 'sort out' what had to be done. When things went wrong – as they frequently did – then someone would have a word with someone else. Perhaps somebody would go down to the site and sort it out. On the other hand, planning of the design process sounded like extra work. Some would even argue that the design process could not be planned.

Even today, this tradition lingers on in many design organisations. Any degree of planning or management of the work process is not conspicuous. Designers tend to be very individualistic. They are professionals and they do not like to be directed by others. Management structure is ill-defined and *ad hoc* decisions are taken as necessary. The work gets done somehow.

62

Many CAD systems are introduced into this sort of environment. The following scenario is fairly common. A CAD system is selected. It is foreseen as something that will boost the firm's prestige, and at the same time it might well extricate the firm from many of its current difficulties. Some accommodation is found for the hardware, the equipment is delivered, and later one or two people spend a few days being trained to operate it. The supplier departs from the scene, and these raw recruits to the world of new technology are left to cope as best they can. Later, perhaps months or years afterwards, a general realisation dawns that this CAD system is costing a lot, yet bringing few benefits. The firm then decides that it has been a victim of overselling of the concept of CAD, and that the old and well-understood ways remain the best. The 'old hands' have given the system a wide berth, because they know that it is unwise to be associated in any way with a failure. The system is doomed, but of course it was doomed right from the start.

I suggest therefore that modern CAD technology does not mix with these old-style design organisations that operate in a management vacuum.

5.2 THE NEED FOR MANAGEMENT IN DESIGN ORGANISATIONS

If management is felt to be needed solely to allow a CAD system to function properly, then perhaps it would be best not to implement a CAD system at all.

Even if we ignore CAD for the moment, there are many reasons why design organisations nowadays must have good design management if they are to survive:

1   At one time the construction industry was slow-moving and backward. However, new techniques, new materials and new construction forms have been introduced during the last two decades. Some of these have not always been handled well. A few random examples include multi-storey public housing, flat roof construction, and fixings of wall cladding panels. The rate of introduction of new methods is increasing markedly now, and better design management is the key to their successful adoption.
2   Projects of all kinds are becoming steadily more complex. More design experts are required to cover the wider spread of expertise needed. Multi-discipline design teams with more members are inevitably more difficult to control and co-ordinate.
3   The construction industry has lost many of its skilled craftsmen. This means that designs today must be documented more carefully and in far more detail. Fewer problems can be left to be sorted out on site.
4   Statutory requirements are rapidly becoming a more serious legal, administrative and technical burden to designers.
5   Clients are more cost-conscious and expect better value for their money. This is reflected in the cost-cutting exercises which are such a feature of the design process now, involving continual changes to the design. At the same time clients are more time-conscious. Design solutions are needed faster.
6   Today there is a new and highly competitive business climate. Not least, this is reflected in fee competition among professional firms. To be successful today, a firm must provide a really useful service to the community at large.

For these and many other reasons, the organisations that harp on about outmoded themes, and which continue to design by 'seat-of-the-pants' methods, are heading for failure. Difficulties might for example appear first in their efforts to seek professional indemnity insurance. Conversely it is the design offices that can respond quickly to new circumstances that will win.

The role of management is essentially to provide direction for an organisation.

In former times when there was less need for changes of direction, little in the way of management was required. The swift responses and rapid changes of direction now required are only possible when there are competent leaders and more emphasis on business ability.

Now we must turn our attention back to the implementation of a CAD system. CAD is a technology which clearly is developing very rapidly. Introducing CAD implies a strategic change to the work pattern of a design office. Management – which I have argued, every design organisation needs anyway for many other reasons – must be in place to provide the direction for this critical change in work pattern. Some organisations are certainly positive in their outlook. They recognise CAD as a potential benefit. They see it also as an heaven-sent opportunity to rethink the management of their whole design approach.

Management, when equipped with CAD design tools, must essentially be concerned with finding and introducing ways in which individual professionals can co-operate so that better results can be consistently achieved within acceptable timescales.

It is essential that attention is given to all these implementation topics starting from a time even before the order is placed with the supplier. However, many aspects of these implementation chapters will be of practical use to those design offices that are already operating a CAD system.

5.3 CAD MANAGEMENT STRUCTURES

First I want to discuss the merits of four different types of CAD management structure that have been adopted in design offices:

(1) Undefined
(2) Centralised
(3) Decentralised
(4) Hybrid

Undefined management structure

Some might argue that this is not a form of management structure at all, but the lack of it. It mirrors the undefined mode of manual design already discussed, where individuals make *ad hoc* arrangements and decisions, the work gets done, but nobody is too clear about how it gets done!

This form may be adopted by default, to reflect what exists prior to the CAD implementation. When the CAD system arrives, a number of people are trained to operate it. Then essentially they co-operate to a greater or lesser extent to undertake the work among them. There are few formalities involved.

This mode of working can work reasonably well, but only in small organisations where everyone knows everyone else and a very strong spirit of co-operation exists. The door to the CAD room usually acts as the control valve. If the team is too busy, the door is closed and they are not available. This is not ideal from the point of view of the rest of the design office.

Even here, someone usually emerges as a key figure or leader. I feel that if this undefined mode of working is to be effective, the team should agree to operate according to many of the principles which I shall be discussing later in this chapter.

Centralised management structure

Here the hardware and software, together with trained operators, are all formally combined into a compact and clearly defined group or work section within the organisation. The group has a leader to manage all its work. Essentially it provides a specialist service for the rest of the organisation. The group operates in some ways like an independent CAD bureau, but is located within the firm itself. Sometimes it might even be treated as a separate profit-centre.

This centralised structure is quite common. It is seen as an efficient means of bringing together people who are prepared to specialise in various aspects of this new technology. With their dedication to the cause, such a group can certainly make the new system run within a short timescale. The group will be keen – in its own way – to provide its service to all the other staff.

The benefits of this approach are bound up with the fact that centralised project control is easier. This approach is particularly appropriate where small CAD systems having one or two workstations are implemented.

The problems with this centralised approach stem from the specialisation itself. All too often, the CAD group does not become well integrated with the rest of the firm. It is remote, perhaps even considered too elitest, because the members talk with their own jargon which nobody else understands. Communication between project designers and the group is difficult. It is not easy for project designers to exert much control so as to get work done within their own all-important project deadlines. When approached to tackle some work, the CAD group always seems to be overloaded with other priorities.

The centralised group will tend to be composed of technicians, rather than project-orientated designers. The danger in this is that it might be able to provide little more than a drafting service. In extreme cases, the poor communications mean that it actually becomes virtually a tracing service for drawings.

I have perhaps painted a poor picture of the centralised approach. It must be said that it can and does work in some circumstances, especially where those concerned are aware of the dangers. For project modelling work however, it seems to me that the cards are usually stacked against this centralised approach.

Decentralised management structure

Here the applications and projects are all-important and the system resources are divided up among all those that will apply them. Project designers can therefore exert control over their own share of the CAD resources, and use them as they see fit.

It is clear that the strong feature of this arrangement is that, potentially, CAD will become better integrated and better understood throughout the organisation. Giving full control to the ultimate users is usually a worthy aim.

Decentralisation depends on the users being properly trained. We shall discuss later that this means much more than teaching a few people to operate the workstations. It is a tall order to expect many individuals, thinly spread throughout the organisation, to become adequately conversant in a short time with these radically new techniques. In effect, the danger with decentralisation is that the real CAD expertise acquired or built up by the organisation will be diffused too much. Then the system cannot be fully effective.

Inevitably people do vary in their ability and enthusiasm. Some will take up the facilities provided and use them; others will not. Yet the resources concerned are too expensive for some of them to be under-utilised or misapplied, or even left idle. In any case, workload fluctuations within the firm can tend to create problems with the utilisation of decentralised resources.

Another disadvantage is that project control is more difficult. One department may not know what another department is doing on the same project at the same time. Valuable experience and more tangible items like reusable CAD modelling components or standard details built up in one department may not be transferrable elsewhere.

Let us now look at the hardware and software. It is certainly easy to disperse a few stand-alone or networkable workstations. With host-computer based systems it is possible to disperse the terminals – as long as they can be connected somehow to the host (we will look at these hardware terms in the next chapter). However the computer itself and peripheral devices such as plotters always have to be centralised somewhere, to be controlled and shared by all.

Decentralisation seems like a good long-term goal (meaning years rather than months), but is not easy to achieve this solution initially.

Hybrid management structure

By 'hybrid', I mean a management structure which exhibits features of both centralisation and of decentralisation. It is a compromise arrangement where we search for the advantages of each approach. Essentially the aim is to marshall certain of the scarce resources of people, equipment and software in such a way that their services are concentrated and shared. For convenience and flexibility, other resources are spread around to where the applications lie within the organisation. The application of the dispersed resources is co-ordinated.

I am suggesting that most organisations should apply some form of hybrid management structure. For this reason I shall be concentrating on this approach in the rest of this chapter. Since the term 'hybrid' encompasses every solution lying between the two extremities of centralisation and decentralisation, there is plenty of choice to suit individual design offices. Indeed, there is no blueprint which can be adopted slavishly.

In establishing a suitable CAD management structure, we must look at the roles of the people who will be most concerned with the implementation and functioning of the new CAD system. The following ideas are put forward as guidelines.

5.4 HIERARCHY OF ROLES

I believe that before the CAD system is introduced, four new 'roles' must be established within the organisation. These are:

- CAD director
- CAD co-ordinator
- Computer manager
- CAD operators

If the organisation has not had a CAD system before, then of course these are completely new roles. These are not necessarily appointments or jobs, because they might not be a full-time occupations. Indeed one individual might cover one or more of the these roles within his job specification. What is important is that each role should be recognised and the responsibilities clearly assigned to people. Preferably this should be organised well before the hardware and software arrive.

Fig. 5.1 shows one way in which these roles might be integrated into an existing design organisation. The arrows indicate the main two-way exchanges of information. We shall see that the link between CAD co-ordinator and project managers is the key relationship.

I shall continue to adopt these particular role names, although other names might be deemed more appropriate as job titles within individual companies.

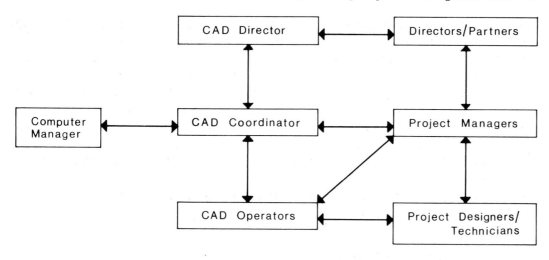

Fig 5.1 CAD organisational structure.

5.5 CAD DIRECTOR

The CAD director should preferably be a director, partner or at least a very senior executive of the company. Clearly this will be a role which the individual concerned must combine with many other responsibilities. Yet this new duty must not become too submerged. Responsibility for changing the long-ingrained practices must not be undertaken lightly.

The following indicates the nature of the responsibilities of the CAD director:

1 Holds the main overall responsibility for CAD use and represents the interests of the highest level of management or owners.
2 Allocates (or appoints) individuals to undertake the roles of CAD co-ordinator and computer manager.
3 Acts as the link between upper management and all the staff who will be involved directly or indirectly with the system. Essentially he must try to create and maintain a sense of unity of purpose.
4 Responsible for the establishment of clear CAD objectives. The system must produce results fairly quickly, and be seen to be doing so. Realistic goals must be established by the director working in liaison with the CAD co-ordinator.
5 Later when the system is operating, monitors the situation to ensure that these objectives are being achieved in practice. Initiates actions if they are not.
6 Ensures that CAD does not become submerged in one department or taken over for one project. Rather it must be applied across departmental, disciplinary and project boundaries.
7 Encourages other people within the company to co-operate in such a way as to make success of the whole operation likely. In particular, supports the CAD co-ordinator in the latter's dealings with his peers in the organisation, who might themselves be the designers responsible for individual projects (project managers).
8 Arranges for resources to be made available when needed. These might be people, space, materials or money. For example, the CAD director would be directly involved if proposals were made to upgrade or enhance the system in the future.

The appointment of a CAD director will demonstrate the commitment and active support from upper management. This is essential if the staff lower down are to be convinced that CAD is not being relegated as a fringe activity in the firm.

The CAD co-ordinator would normally report directly to the CAD director. I see this director therefore as being a high level ally of the CAD co-ordinator. He frees resources, gives encouragement when and where needed, and fights the occasional top-level battle against ignorance of what the computer is intended for.

Clearly it is essential that the CAD director has a good overall awareness of at least the principals on which the system operates. Although he may not need to have an intricate knowledge, nevertheless this person must allocate adequate time for his own awareness training.

In Chapter 4, I recommended that a member of the firm's upper management should head up the team to undertake a CAD feasibility study. It would be ideal if this same individual could retain his interest and move on to become the CAD director. This person, having recommended the system in the first place, would then have the strongest possible interest in seeing that it succeeds in practice.

5.6 CAD CO-ORDINATOR

has WHFP a cad orot ?

I regard this as the key CAD role within the organisation. It may or may not be a full-time occupation depending on the size of the system but no CAD system should be implemented without a CAD co-ordinator.

The CAD co-ordinator has overall control of how the system is used and what it is used for. Essentially he directs the CAD resources in the execution of project work. The role is directed towards people management and is project orientated. This contrasts with the role of the computer manager who, as we shall see, has responsibility for the continued availability of the hardware, software and data.

The CAD co-ordinator will embody most of the real CAD experience of the organisation. Such experience will initially be in very short supply. It is precisely because of this that we must create the right conditions for the experience to be developed and concentrated in the one individual. With the available CAD experience focussed in this person, the individual must undertake a high profile co-ordinating role. The key function is to co-operate with the various project managers throughout the organisation. He must be able to move about within the organisation, actively liaising with senior and junior people irrespective of department or discipline.

The CAD co-ordinator is the person who must 'champion' the introduction of the new ideas and techniques. Clearly he must be enthusiastic, yet realistic, about CAD in principle.

For this reason, it would be ideal if one of the design office representatives on the feasibility team – discussed in Chapter 4 – could emerge as the CAD co-ordinator. Then he would have been involved in the planning of the system from the start. He would also be committed to the choice of the proprietary system acquired, and so would have plenty of incentive for making it work in practice. If this is not possible, the CAD co-ordinator should be appointed as soon as possible, and preferably well in advance of delivery of the hardware and software.

Obviously it would be a bonus if the CAD co-ordinator had had some previous experience of CAD, if only as an operator. In most cases this may not be feasible, and lack of previous CAD experience should certainly not be a bar. After all, most organisations already operating CAD had to start with little or no past experience. Nowadays, management must be ready to find a suitable candidate and make provisions for appropriate initial and on-going training in CAD. Important qualifications (proven or potential) are:

- Design management experience is the main attribute.
- An ability to organise and manage others.
- A willingness to co-operate with other people.
- A good communicator.
- CAD application knowledge.
- A basic enthusiasm.
- A willingness to learn.

It is better if the co-ordinator comes from a professional design and project management background – rather than being a technician. Certainly it is design management experience, and not a computer background, that is required. Deficiencies must be countered by training.

In its selection, management should show a strong preference for finding a person of suitable calibre from within the firm. This is because a knowledge of the work patterns of the company, and of the people working in it, are valuable assets. Only when the company has nobody suitable or willing to undertake the role, should it consider recruiting externally.

In all cases, management – actually the CAD director – must arrange for conditions to be as conducive as possible for success, especially in the early stages of the venture. If there is nobody to shoulder this role of CAD co-ordinator, it is almost inevitable that the use of the CAD system will be fragmented, directionless and ineffective. There will be duplication of effort on the part of operators and others. Many of the benefits from CAD will not be realised.

A brief outline of some of the main duties and responsibilities of the CAD co-ordinator is as follows:

1 Gaining familiarity with the system, so as to become the principal expert and advisor on it within the design office.
2 Planning the siting and accommodation of the equipment (in conjunction with the computer manager).
3 Setting up and maintaining a pool of people who are able to operate the system.
4 Organising the necessary training and awareness. Later when the co-ordinator has become well acquainted with the system, he may undertake much of this training personally.
5 Liaising with project managers/designers elsewhere within the design office to identify the work which is suitable for the CAD system. This is crucial during a period of rapid change.
6 Programming and co-ordinating the flow of work undertaken with the system. Determining work priorities.
7 In liaison with the CAD director, establishing realistic goals or objectives with which system achievements may be compared.
8 Monitoring work done, to check on system achievements.
9 Setting up company-wide and project-wide classification systems for design information. By this I mean category and layering systems, or their equivalents (see Section 10.5). Setting up naming conventions for components.
10 Liaising with operators and designers to save selected components, symbols, details, and drawings in system libraries so as to be available for future use.

11   Establishing and maintaining house standards for the project documentation produced with the system.
12   Fostering a team spirit and enthusiasm among users.
13   Acting as a focal point for queries and problems when they occur, but also trying to foresee and avoid problems developing.
14   Liaising with the user-group. This is the association of design companies that individually operate the same proprietary CAD system. Normally the computer manager will handle the contacts with the system supplier.
15   Sales, marketing and public relations. Even professional design firms need to give attention to these matters. Most organisations soon find that the CAD system has an important role to play here.

It may be appreciated now that this is quite a demanding role. The choice of CAD co-ordinator can be far more important than choice of the hardware/software. As a result it should receive a lot of attention from the CAD director. The chosen individual should receive as much understanding and support as possible from the rest of the organisation.

Some architects and engineers see this essentially as a technical role which they regard as beneath them. This is short-sighted. In reality the appointment is usually much sought after, and rightly so. It is certainly a new role, and many people wonder how their career prospects would be affected if they were to undertake it. It is true that the co-ordinator will not get the experience and satisfaction of managing a project and seeing it through its various design stages. However there are already a fair number of people fulfilling the CAD co-ordinator role, and all indications are that their career prospects are very much enhanced. They are indeed placing themselves in an excellent position to gain management experience of a kind that will become increasingly valuable in the future.

The co-ordinator role as described might well be a permanent and full-time job for an individual in a design office where a large CAD system is being implemented. Indeed a large system with sophisticated applications or projects will demand a greater degree of co-ordination and management resources. In less demanding circumstances, the job might be full-time for a period of say six to twelve months. After that it might reduce to become part-time. Then the person could take on some parallel duties such as managing a suitable (i.e. not too demanding) project. Initially it would be better to keep to as flexible an arrangement as possible and see how it works out in practice.

Back-up for the CAD co-ordinator

Management must nevertheless avoid becoming over-dependent on this chosen CAD co-ordinator. This is a danger simply because it is such a key role. To achieve a measure of independence, management must in addition identify a leading operator, or someone else in the organisation, to take on the function of Deputy to the co-ordinator or lead operator. This person will take over the co-ordinator's role when he is on holiday or ill. Also when a shift system is in operation, the deputy can cover when the co-ordinator is not present. The company will then be much less vulnerable if the co-ordinator decides at any time to resign.

5.7 COMPUTER MANAGER

Whereas the CAD co-ordinator is primarily concerned with the application of the system, with training, users and with the project work done, the computer manager role is more technical in nature. Here lies the responsibility for the various items of computer equipment, the software, computer data and for the smooth running and maintenance of all of these.

Where there are other types of computer operations within the company, there may already be a computer manager, or data processing manager. This person could perhaps shoulder the added duties of looking after the CAD hardware and software.

In a big organisation, the computer manager role would tend to be a full-time appointment. On the other hand, where a ready-to-use CAD system of moderate size is installed and few other computer applications exist, then this computer manager role could be undertaken by one of the CAD workstation operators. The person concerned would of course require some appropriate training.

A fuller explanation of the computer manager's role will be emerge in Chapter 6 where I concentrate on the hardware, software and data aspects of implementation.

I have been at pains to separate out the roles of CAD co-ordinator and computer manager so that they may be well understood. It might be possible for one person to tackle both, and this is sometimes done. However it is clear that the co-ordinator role is a new and demanding function on its own, and adding the additional computer responsibilities might not be prudent. The roles can and often will be split between two people. The computer manager could report directly to the CAD co-ordinator. Alternatively the two could co-operate as peers in the organisation, reporting individually to the CAD director. We should not have the situation however where the CAD co-ordinator reports to the computer manager.

5.8 SELECTION OF OPERATORS

It is the operators of the workstations who actually will drive the system and so it is very important to single out who these people will be. A nucleus of operators has to be chosen by management at an early stage, and certainly before the system is delivered.

Management is often mystified about the kind of person who should operate. There are three main options:

- technicians or draftspersons,
- designers; engineers or architects,
- technically unskilled people.

We must look at the people available, at the nature of the workload, and the type of CAD system acquired. There may well be merit in choosing a balanced team of individuals drawn from more than one of these categories. When selecting operators, there are some basic considerations, as follows:

- The best tools, i.e. CAD, should in principle be given to the most capable people – in whatever category they lie. It is better to improve the efficiency of already effective people, than to try to make some improvement to poor workers.
- CAD is still a scarce and expensive resource. For this reason, it is important to select people who will treat it seriously and become fairly dedicated operators. People who would use it on a rather casual or intermittent basis are probably not so suitable.
- Learning how to operate the system may take only a few days. Learning how to get the most out of it and to use it really effectively requires months or years of experience. This is true of all CAD systems except perhaps those that are so limited in scope that their use would bring few benefits anyway.
- *Technician operators*: The majority of CAD operators today are technicians. This is

because current systems are predominately for drafting. Technicians are better able to specialise in CAD operation, and to become dedicated and efficient operators.

- *Professional designers as operators*:   Designers have so many and varied duties and interests that they can find it difficult to dedicate themselves enough to operating. Compared to technicians, they tend not to become as proficient and speedy at basic operating. Of course there are exceptions, and there are compensating factors.

  Where their design skills can be utilised, these can compensate for some inefficiencies in operation. For example, designers may be better operators than technicians where design work is being performed using a computer modelling system. This is because the designer's ability to engage in creative work by directly operating the system could well pay dividends. The alternative, when the designer works 'at arm's length' through a technician operator, extends the communication lines. The distinction between efficiency and effectiveness needs to be drawn here.

  If the system is to be used for preliminary design of projects then, as will be discussed in Chapter 7, this work is often undertaken by high-level designers. This means that one or two of these may need to be trained. There is a special problem here because designers may use the system very intermittently and so can be expected to acquire only limited operating skills.

  The trend towards designers doing more of the operating is likely to continue as reductions in workstation prices and improvements in design software work their way through.

- *Technically unskilled people*:   People must know what they are drawing or designing. This applies whether they are using a computer aided system or working on the drawing board. The scope for using technically unskilled people in CAD operation therefore seems rather limited. Occasionally such people have been used with success when given very close supervision and employed on large and highly repetitive projects. In other situations, the CAD work could be in danger of descending to the equivalent of mere tracing. There is little point in using an advanced system for this limited purpose.

- There seems to be a widely held view that young people make better operators. However there does not seem to be any rational basis for this assumption. As discussed previously, the main handicap of older people is their own tendency to assume that they are at some disadvantage. In practice their advantages are their experience and knowledge of the design discipline. This is something which is far more important than a temporary lack of CAD operating skill. Background, initiative and attitudes are more important than age.

- Clearly the people chosen must reflect a cross-section of the disciplines, applications, sections or departments, and projects for which the system is intended.

A decision has to be made by upper management as to whether all operators will report directly to the CAD co-ordinator as indicated in Fig. 5.1. An alternative is for them to continue under the various section or project leaders. Individual circumstances, personalities and preferences must dictate this choice. One factor is the capability of the CAD co-ordinator as a manager. In many circumstances the best approach is a composite approach:

(1)   to decentralise applications of the CAD system by allowing section or project heads to retain technical, project and administrative control of the operators and their work.

At the same time . . .

(2) to centralise or co-ordinate the computer operations, including access to the workstations, mode of operation, best use of system, etc., by allowing the CAD co-ordinator to exercise control over the operators in these respects.

This is a 'matrix' form of management control.

It is wise to discuss the introduction of CAD and the selection of operators with all the technical staff. This should be done early, preferably well before a system is even ordered. The aims and objectives of management should be discussed frankly and openly. This will go some way at least to alleviate fears and prejudices forming. In large organisations or public authorities, the staff associations or unions ought to be consulted, and the same principles apply. In all cases much care must be exercised in the selection process. Avoid showing illogical favouritism. Volunteers, enthusiasts or informal leaders should qualify for some degree of preference.

In general, few people seem to turn down the chance to acquire the new skills. The opportunity can lead to job enrichment. There is no need for management to immediately reward with larger salaries the people it has selected for operating. The proper time for considerating such a course is when the individuals concerned have acquired the skills and have amply demonstrated an aptitude and willingness in deploying them.

Management sometimes feels that it is important to identify as many people as possible to train as operators. It is probably wrong to do this initially. Around two or three operators per workstation is likely to be adequate at the start.

5.9 TRAINING OF CAD CO-ORDINATOR AND OPERATORS

Training is essential

Whereas there is a new realism that money has to be spent on hardware and software to remain competitive, there seems to be less support for investment in training people. Especially in the construction industry, the record on investment in updating skills is very poor. Training must be taken seriously and furthermore the cost and time required for it should not be under-estimated. People do not learn to draw on a drawing board in a day or so, and the same is true when they want to draw or design with the aid of a computer.

Normally the initial training of a batch of operators will be included within the system supply contract. It is carried out by the vendor's own staff. However, a much wider view of CAD training than this is advisable.

Training of the CAD co-ordinator

First we must consider the CAD co-ordinator. Clearly he must be able to operate the system at least as well as the other operators. In addition, he must quickly become totally acquainted with the system and with the design management issues involved. His training should precede that of the operators, so that he can get ahead – and be seen to be ahead. Then he can begin to fulfil a role of adviser and problem solver almost from the outset.

Most system vendors give very little attention to training in system management. Perhaps all that is done is to single out the person for some special attention during the initial operator training sessions. This is not enough.

Ideally his training must start immediately the order for the system is signed. This means that it must be done externally to the company. Perhaps arrangements can be made for him to attend the operator training for some other customer that is a month or two ahead in its acquisition procedure. Perhaps he can spend a week or two with a firm that has at least a year or two of operating experience with the same system. There is the danger in this course that he picks up all the faults of that organisation. Some time spent with a consultant could be valuable in tailoring the training and making it more company-specific. **t²** Solutions Ltd has an associated company called T Three which specialises in providing this CAD management and application support. One or more of these ideas should be implemented because we have seen that so much will depend on this co-ordinator.

After his initial training, the co-ordinator needs plenty of access to a workstation as soon as the system is implemented.

The initial training and experience for operators

The initial trainees are the ambassadors as far as the firm's use of CAD is concerned. If this initial training is botched they will not get an adequate grounding and will be unable to use the system properly. This will limit the effectiveness of the whole expensive CAD implementation. By contrast, good training will maximise the value of the system, and this is why so much attention needs to be focussed on all aspects of the problem.

As far as the location and timing of the training for the initial batch of operators is concerned, there are two possibilities:

1 *Off-site*: Held immediately before the firm's system is delivered. The training would normally be held at the vendor's premises. With accommodation, subsistence and hardware costs added, this might seem to be an expensive option. On closer examination it may prove worthwhile because the company will not be faced with an initially unproductive system when it is delivered. External training ensures that the trainees are removed from their normal work environment. They are less likely to be subjected to interruptions and so can concentrate totally on their training.

2 *In-house*: Held immediately after the new system has been delivered. The training can be a little less intensive, leaving more time for the course material to sink in. It can be organised so that plenty of practice is gained at the workstations between formal sessions.

Having chosen the operators and arranged the off-site or in-house training, management must make sure that the trainees are seconded properly, despite any competing short-term pressures. This is never easy because the work of the firm must continue and these trainees are often busy people. The CAD director has a role to play in emphasising this point with the trainees' superiors.

The trainees must be receptive to the coming changes in their work practices. Such training is a rare opportunity for them which they should not fail to grasp.

The vendor's training course should be a carefully structured programme. The tutors may have repeated it often, with different customers, so they ought to be competent. Ideally they should be people of the same discipline as the trainees, so that empathy can be established between them. Sometimes the tutors are computer-orientated and prone to the use of computer jargon. Such jargon should be avoided whenever possible and explained carefully where it is really useful. A set of structured notes containing background material is always useful.

Time must be set aside for practice. This too should be structured to cover all relevant aspects of the system operation. Trainees soon tire of merely 'playing' with the system and they must on no account be allowed to become bored. Tutors should be present for at least part of the time, and trainee progress monitored.

After training, the new operators should move on as soon as possible to work on real projects. This way confidence and competency can be built up quickly. This is where the early start made by the CAD co-ordinator comes in useful, because he can plan and direct this initial work. He should be available during this time to cope with difficulties and queries.

Pressure should not be put on the new operators too soon after their training, or too much expected from them. They may be competent designers or technicians in their own right, but with the new computer tools it is best to treat them for a time as if they are raw recruits. Putting a new operator on to a tight production schedule is certainly a recipe for disaster.

Another form of pressure which must be avoided is to ask a new operator to provide *ad hoc* demonstrations of the system for upper management or visiting VIPs. Even experienced operators can find this a tough assignment. Of course demonstrations will have to be given before long. It is best if a few routine operations can be practised for this purpose. Also the operators should retain some examples of their work which they can call up on the screen on demand. Departures from well-prepared routines should be avoided at least in the early days.

Few new operators experience any major difficulties which cannot be overcome. They tend to become keener and keener as time passes. Nevertheless, it would be a mistake to expect all the new operators to reach an identical level of competence. Indeed the system can be more flexible in operation if a range of skill levels becomes available.

If however someone proves not to have an aptitude for operation, then this must be recognised and he must be replaced. The training already provided to the individual will not have been worthless, because he will at least have gained a useful insight into CAD methods. He might well be able to work very effectively in conjunction with a CAD operator on future project work.

On-going operator training

After a few weeks the training and initial practice will gradually be transformed into serious project work. Build-up of speed will take several weeks or months of further work. Management should make sure that each operator gets access to and works on the system regularly, because otherwise some of their skills may become eroded.

In a sense, CAD education will never stop. Indeed it is to some extent self-perpetuating because operators will learn new ideas, work-arounds and tricks from each other. The CAD co-ordinator or the leading operators should try to fulfil a continuing role in directing and encouraging this. In larger organisations where personal communications might not be so easy, it can be useful if regular in-house user meetings can be held so that people can air their difficulties and swop ideas.

Refresher courses or advanced training can be organised as necessary. These are best conducted in-house because then they can be company-specific. However some input either from the vendor or from training consultants can prevent the participants from becoming too insular in their attitudes and practices.

System upgrades have to be absorbed and these often require training at the appropriate time.

When the system is in full productive use, there will come times when additional operators are needed and have to be trained. It would be possible to arrange more training with the system vendor of course. However, by this time it may be preferable to undertake the training in-house where it can be more specific to the project work in hand. The CAD co-ordinator or one of the by-now experienced operators would then act as the tutor.

All in-house training will lead to some loss in productive system resources, and this must be planned and allowed for. One authority[14] has found that the initial and on-going training required 12% of the available workstation time during the first year of operation.

Computer manager

I have not yet mentioned the computer manager's training. Whoever fulfils this role will need some specific training in the management of the hardware and software. This will deal with such matters like starting and stopping the system, running the plotter, disc back-ups and data archiving, and managing the consumable materials. These matters will be discussed in Chapter 6.

The system supplier's firm is usually computer-orientated and has plenty of experience in running its own computer system. It is usually well able to provide specific training for the computer manager. However it may have to be prompted to do this.

5.10 TRAINING THE ORGANISATION GENERALLY

Awareness training for CAD director and project designers

This is all important, yet surprisingly it is often completely neglected. Many people in the firm will become users of the system even though their use is at arm's length through operators and not by using the workstations directly. If there is little or no awareness of the CAD potential and mode of application within the organisation generally, the CAD system will only function at best like a remote bureau operation.

Apart from a special case (the CAD director) the key individuals are the managers or designers who are responsible for design projects. Section or department leaders are also important. These people cannot be left in the dark about the CAD system if it is going to be used effectively in the office.

CAD awareness training is providing to people, who are not literally hands-on operators, a good appreciation of the principles on which the system runs. A realistic view of the system's capabilities and limitations must be imparted to them. They do not need to know all the operating details and commands of course, so their training is quite distinct from operator training. Awareness training is also needed to prevent prejudices, misconceptions and needless barriers forming between the CAD operators and the others.

Whereas the number of operators selected for training should be related to the number of workstations, awareness training should be widely available. To keep costs down in these circumstances, some formal lectures, demonstrations and discussion sessions could be arranged and use could be made of videos.

Some team leaders might profit from some low-grade operating knowledge. This would enable them to display on workstations or even to plot the project data created by their assistants. This would give them added confidence and an ability to retain a better control of their design projects.

In addition to specific awareness training, the CAD co-ordinator must strive at every available opportunity to raise the awareness level among his colleagues.

5.11 WORKSTATION SESSIONS

Work the equipment hard!

It is an unfortunate fact that the CAD workstation remains a scarce and expensive resource. The equipment has a high capital cost and low running charges. This means that the hourly cost of use becomes steadily lower as the number of hours of productive use per week increases. Looked at another way, the marginal cost of each additional hour of use within a week is very low. Like all such capital-intensive resources we must organise ourselves so as to make the best possible of it, even if this means some inconvenience to operators and others.

As the years pass, workstations are likely to become cheaper while operators' salaries will continue to inflate. Eventually a time may come when workstations are considered to be cheap, perhaps even as 'part of the furniture'. Then everybody can have one on his desk to suit his convenience. We must plan the work not for then however, but for today's circumstances.

No operator should be allowed to acquire a workstation and retain it solely for his own use. This does happen in some design offices for it is extremely convenient for the person in question. However operation of a CAD workstation is, or should be, very intensive work. An operator's efficiency inevitably falls off after three or four hours of continuous work. Use by one operator throughout the day would therefore mean that it is being used inefficiently throughout much of the time. Worse still, it would be idle during breaks for meals, refreshments and other inevitable interruptions. The time required to seek and prepare information off-system can be at least 25–50% of the time required for input at the workstation. Then there are always many other office activities that any person inevitable is caught up in. As a result it is better to undertake CAD operation in a series of structured sessions at a shared workstation.

Draftspersons are likely to use the equipment more frequently or for longer periods compared with designers.

The principle of sharing stations creates contention problems which must be minimised. When a station is not available to an operator, he cannot turn to the drawing board. Unexpected delays are wasteful. So work rotas must be drawn up for each station to ensure co-ordination, co-operation and flexibility.

Where there is a high-priority project, perhaps one or more stations might be dedicated to it.

The number of trained operators is critical. If there are too many, there will be contention for workstations and manpower will be under-utilised. The degree of contention can be reduced either by extending the working day or by enlarging the CAD system. If there are too few operators, there may be little contention but the computer system will be under-utilised. A dynamic balance must be maintained between:

(1)   number of operators,
(2)   number of workstations available,
(3)   project work in hand,
(4)   daily pattern of work.

Flexible working and shift working

Many different arrangements are being used. For example the workstation can be operated in:

- two 4-hour work-sessions within a normal day (extending uninterrupted through the lunch period);

- three 3-hour sessions within a slightly extended day;
- three 4-hour sessions.

Some organisations have even operated throughout the night, at least when the work-load peaks have occurred on particular projects. In principle, the most routine of operations and the most computer-intensive work should be reserved for nighttime.

There are special procedures that will need to be handled routinely at a time when the workstations are not likely to be much in demand. These include dis~-backups, data archiving, and large plot runs. It is very useful if one member of staff can be found who is willing to work regularly on these procedures outside the usual hours of system usage.

Extending the working day for the equipment of course means that some form of flexitime or shift working system has to be instituted for the operators. It is important that some understandings are reached on this during the operator selection stage. If management's aims are frankly and openly discussed with the candidates, usually some reasonably amicable arrangements can be made. Clearly the work patterns must be set so that those operators working on early and late shifts have enough off-system 'core' time to enable them to communicate with colleagues working on the same project.

Workstation rotas and sessions

The organisation of such rotas should become the responsibility of either the computer manager or a senior operator. Responsibility here involves setting up a scheme for rotas, ensuring that it runs smoothly, and the resolution of problems. The individual operators should have as much freedom as possible to organise themselves.

Ideally the team must settle down to an effective pattern of working. In practice it is not easy to maintain a uniform pattern in the face of project pressures and unexpected events of all kinds. The co-ordinator must nevertheless try to instil a sense of priority in system operation. Rotas for access to stations must be set up for the weeks ahead. While no changes should be allowed for frivolous reasons, nevertheless a lot of flexibility must be tolerated where the reasons for changes are good.

Project leaders must be made aware of the undesirablility of changing set rotas. So often in an organisation, project priorities are arbitrarily upset simply because too little forward planning is being done within the company.

Operators must be punctual, both in starting and stopping their sessions. Many will find this amount of discipline to be irritating at first. But it must be imposed (perhaps with the help of an alarm clock!) and it will soon become almost second-nature.

Ideally an operator would have one session on the system per day. This would be followed by a period spent at his desk. The desk-time must be used, at least in part, to prepare for his next workstation session. If the operator is a technician, this desk-time is when he may need to communicate with the designer involved.

Before starting each session, the operator must have a clear idea of the work process that has to be undertaken with the system. Then the thinking time at the station – as opposed to operating time – is reduced. For example, he should already know whether suitable components or symbols for his immediate needs already exist within a computer library. If indeed they do, he must know the names by which they can be retrieved, the method of access and how the standard library components have to be modified. Free-hand sketches or notes might be produced and suitable dimensions figured out before the work-session is started.

An operator should be shielded from interruptions like telephone calls when using the computer. Messages can be taken for him and he can call back. His desk-time is the proper time when these can be dealt with. Similarly, no workstation should lie idle while someone has a coffee break or lunch.

In practice it is not too easy to achieve this level of operation. For this reason, the CAD co-ordinator and computer manager need the sympathy, understanding and co-operation from all concerned. A team of enthusiastic and contented operators will make their lives much easier.

5.12 CAD OBJECTIVES AND MONITORING

The CAD Director and CAD co-ordinator should collaborate to set targets.
Some of these will be connected with the implementation of the system itself, e.g.

Date by which the system is fully operational.
Date by which a specific number of operators are trained.

Some targets will relate to project work, e.g.

Manhours required for a project design.
Date by which documentation for a project is completed.

Such targets should be set so that there is a reasonable chance that they will be met, yet they should stretch the staff.

The reason for investing in a CAD system is of course to derive benefits which collectively will become greater that the total cost. As already discussed it is never simple to measure the potential benefits such as increased productivity, improved quality, or better dimensional control. Yet ways must be found. Analysis of the manhours expended on projects designed using CAD and on others designed manually will provide a measure of the increased productivity achieved. Perhaps an assessment of the number of drawing revisions necessary during the site construction phase of a project will throw some light on the design and documentation quality.

The CAD system must start producing benefits as soon as possible. It is far better if the benefits can be demonstrated in some indisputable fashion, and this is why the setting of targets and monitoring is important. Whatever is monitored or achieved should be consistent with the nature of the long-range requirements of upper management.

When systems are demonstratively bringing benefits, it is much more likely that the resources will be found when necessary for an enhancement or upgrade. The converse is also true.

CHECKLIST FOR IMPLEMENTATION – MANAGEMENT ISSUES

1 Upper management to decide on suitable management structure for the CAD system.

2 Upper management to demonstrate its commitment to the need for CAD system management by identifying persons to fulfil the following roles:
   (a) CAD director
   (b) CAD co-ordinator
   (c) computer manager

3 Upper management to decide whether operators will report directly to CAD co-ordinator or to departmental/project leaders.

4 Institute initial training of CAD co-ordinator.

5 Institute initial training of computer manager.

6 Identify CAD operators.

7 Establish the normal daily pattern for work-sessions and flexible or shift working if appropriate.

8 Operators to receive training either off-site before the system arrives or in-house immediately after it arrives.

9 Identify a leading operator or other person who can act as a back-up person to the CAD co-ordinator.

10 Organise awareness training sessions for others in the design office.

11 Choose initial project work (see also Section 8.6).

12 Set up scheme for workstation rotas.

13 Set objectives and a means for monitoring.

CHAPTER 6

Implementing a System – Hardware and Software

This chapter deals with the implementation of the hardware and software. First there is a brief explanation of the various CAD hardware elements and the manner in which these are configured into practical CAD systems. It will become more common for each user to sit at a workstation which includes its own powerful computer. Several of these workstations will however be linked into a data communications network. This is the concept of the distributed system. It is a configuration which seems likely to predominate in the future because it can provide management with far more flexibility. The network permits several users to share design data and provides access for users to shared equipment like a plotter or large disc store.

The accommodation of computers and plotters requires some planning. The workstations may either be concentrated in one area or may be dispersed around the design offices. The merits of each alternative are discussed. It pays if the workstations can be combined into a good working environment. Lighting, heating, seating, furniture and fittings, and wiring are all dealt with.

Important procedures like disc back-ups, archiving of project data and disaster contingency planning are covered. Plotter management receives attention. Data exchange between organisations is becoming increasingly necessary yet this is an area where difficulties abound. This chapter then tackles subjects like system maintenance, logging of computer resources and system upgrades.

6.1 INTRODUCTION

One difficulty in this chapter is that I have to cover all classes of CAD system. Today the range can extend from a single-user system on a personal computer (PC), to a high-cost system for multiple users based on one or more mainframe computers. I will however be concentrating on the principles involved in implementing the hardware and software. Fortunately these principles are similar whatever proprietary system has been acquired. The information given in this chapter can be supplemented where necessary with some system-specific information obtainable from the supplier.

I discussed briefly the role of computer manager in Chapter 5. This function relates to the management of the hardware, software and data, as distinct to the application of these tools in design-related work.

I stress again that this is a role and is not necessarily a full-time appointment for someone. The role of computer manager might either be undertaken by a special appointee, part-time by the CAD co-ordinator, by an existing data processing manager, or by one of the CAD operators. The computer manager's training and duties relating to work-session rotas were also discussed in Chapter 5, but throughout this chapter we will examine many other duties and responsibilities.

6.2 CAD EQUIPMENT

The hardware or equipment in a typical CAD system will include the following elements:

1   One or more *computer processors*, together with memory, disc storage, and perhaps a magnetic tape device for long-term data storage. A processor could be a mainframe, minicomputer, 32-bit supermicro or personal computer.
2   One or more *workstations* for use by operators. Each workstation includes:
   (a)   A display screen with adequately high resolution, and probably colour, so that it can function as a graphics screen. Screens usually are of the raster type, i.e. the image is composed from many closely-spaced dots or picture elements, called *pixels*.
   (b)   An additional lower-quality screen might be provided for the display of text only (such as prompts, messages, menus and to cater for the CAD management function). Sometimes this screen is omitted and then all such text must be accommodated with the graphics on the single screen.
   (c)   Two or more types of input devices. One of these will be the familiar keyboard. Others might be a data tablet, digitiser, mouse, or (less commonly) a joystick, light-pen, button-box or trackball.
   (d)   A desk or table for the workstation equipment to rest on, with additional worktop space for reference material. Also a chair for the operator.
   The workstation equipment must be combined into a good ergonomic arrangement, as will be discussed in Section 6.6.
3   *Shared devices*
   (a)   A plotter for output of the master copies of the drawing-sheets. Normally a plotter capable of producing A0 format sheets is required, although occasionally an A1 or smaller size is acceptable. Small CAD systems generally have a pen-plotter, while large implementations may include an electrostatic-plotter. Emerging plotter technologies include ink-jet and thermal-transfer.
   (b)   A printer for output of text such as schedules and for making screen 'dumps'. The latter are copies made when required of the graphics and text displayed on the graphics screen. This printer is normally A4 or A3 format. Dot-matrix serial-printers are common but better page-printing technologies are coming in, such as laser, ink-jet, thermal-transfer, LED and ion-deposition printers.
   (c)   A page-scanner for direct input of drawings and text. This type of device may be included within an increasing number of CAD systems.

6.3 HARDWARE CONFIGURATIONS

These hardware devices can be linked together in a seemingly bewildering number of different configurations. The numerous arrangements arise first of all because there are so many types of device available for the system developers to pick from. By picking hardware options allowed by the original developers of the particular CAD system, a supplier can usually design a configuration of hardware specially to meet the specific needs of one customer. We shall briefly look at the most common types of configuration under the headings:

(1)   Host systems
(2)   Single-user graphics workstations
(3)   Personal computer-based systems
(4)   Distributed systems

(1) Host system

Here a central computer acts as the 'host' processor for a number of work-terminals. These are all connected directly to it and can be operating simultaneously. The computer is usually a minicomputer, or perhaps for very large companies, a mainframe.

Sometimes the 'computer' actually consists of a cluster of minicomputers or it might even be several geographically-separated computers with data links established between them.

Each work-terminal in a host system normally contains one or more built-in microprocessors and memory. These provide some processing capability locally within the terminal itself, so it is not totally reliant on the host machine.

Acting in a way that is transparent to the user, the terminal may receive and store information from the host computer to enable the local processor in the terminal to generate and display the image on the screen. However the balance between the processing performed within the terminal and the host varies. For example, some systems may actually transfer some part of the project model from the host to the terminal. The user can then work with this sub-model locally within the terminal. Later, the modified version of the sub-model is returned for storage within the host.

Work-terminals are directly connected to the host by multi-core cables, although remote terminals could be connected through some form of high-speed data link such as a leased telecom line.

(2) Single-user graphics workstation

The graphics workstation is typically a high-performance unit which has been specially designed for graphics or CAD work. It is powered by a supermicro processor. Typically this could be a 32-bit processor like the Motorola 68020, Intel 80386, or one of the newer RISC chips, all of which are capable of running with virtual memory under the UNIX operating system. A specialised graphics workstation like this is not to be confused with a personal computer that has been enhanced in some way to give it some graphics capability (see item 3 below).

- several megabytes of memory,
- hard disc store,
- floppy disc or some form of magnetic tape for long-term storage,
- large, high-resolution screen,
- keyboard,
- data-tablet or mouse,
- features like multiple windows, pop-up menus and icon-driven system software,
- CAD software.

This workstation can function as a nearly-complete single-user CAD system. Such a workstation is not automatically supplied with a plotter or printer, so these items will have to be obtained and linked to it to make up the complete CAD system.

(3) Personal computer-based system

Compared with the graphics workstation, the basic PC will only be capable of rather limited graphics. So it must be enhanced in several ways to make it more suitable for computer drafting or for simple modelling work. This can be done, by adding the following items:

- graphics card, to improve the graphics capability;
- co-processor to enable it to cope better with the fairly high precision which we require, this being of the order of 1 mm in a project measuring up to a kilometre or more;
- higher resolution monitor to supplement or replace the normal PC screen;
- data-tablet or mouse to supplement the keyboard;
- CAD software;
- printer;
- plotter – we certainly must not forget this item because it may cost more that the rest of the system put together.

When the PC system is plotting a drawing, it may not be possible to use it for drawing input or any other task. Some plotters allow the data to be down-loaded from the PC fairly quickly into the plotter's own memory. After this, the plotter can operate by itself, leaving the PC free for other work.

(4) Distributed system

When several users of graphics workstations (or PCs) have to work together and share design data, then it becomes worthwhile to set up data-links between them.

Where the workstations are all on the same site, this is done by establishing a Local Area Network (LAN), either baseband or broadband, or else by using telphone-like PABXs.

Remote workstations can be attached through fixed telecommunications links. Shared devices like a plotter or printer can also be attached to the LAN or data-link. An example of what is possible is shown in diagrammatic form in Fig. 6.1. Grindley has discussed computer networks and the beneficial effects on data communications.[15]

The inherent advantages of the distributed system stem from its flexibility. In practice, the individual workstations or *nodes* could be low-performance PCs, mid- or high-performance graphics workstations. Therefore the functionality can be arranged to suit the applications expected at any location. Each workstation could contain its own disc or it could be a disc-less node which relies on the storage capacity available elsewhere in the network.

A large disc store could be incorporated as shown in Fig. 6.1, intended perhaps for the large-scale storage of project information. A *power-node* could be incorporated in the network – this being either a specially powerful workstation, a mini-computer, or a mainframe in the largest companies. Such power-nodes can provide the adequate power which any user may require from time to time, for example for any large computation such as finite element analysis, or a building energy simulation. The power-node can also cope with the system housekeeping function which becomes increasingly important as the number of users in the network increases. Likewise, the power-node can be the base where the database management function operates (refer Chapter 11).

The distributed hardware solution is appropriate even when several designers are working on one large project. In this case, however, it is vital that the design information – the project database itself – is concentrated in one location, i.e. in one disc store to which all the users have access. Distribute the hardware by all means, but not the database.

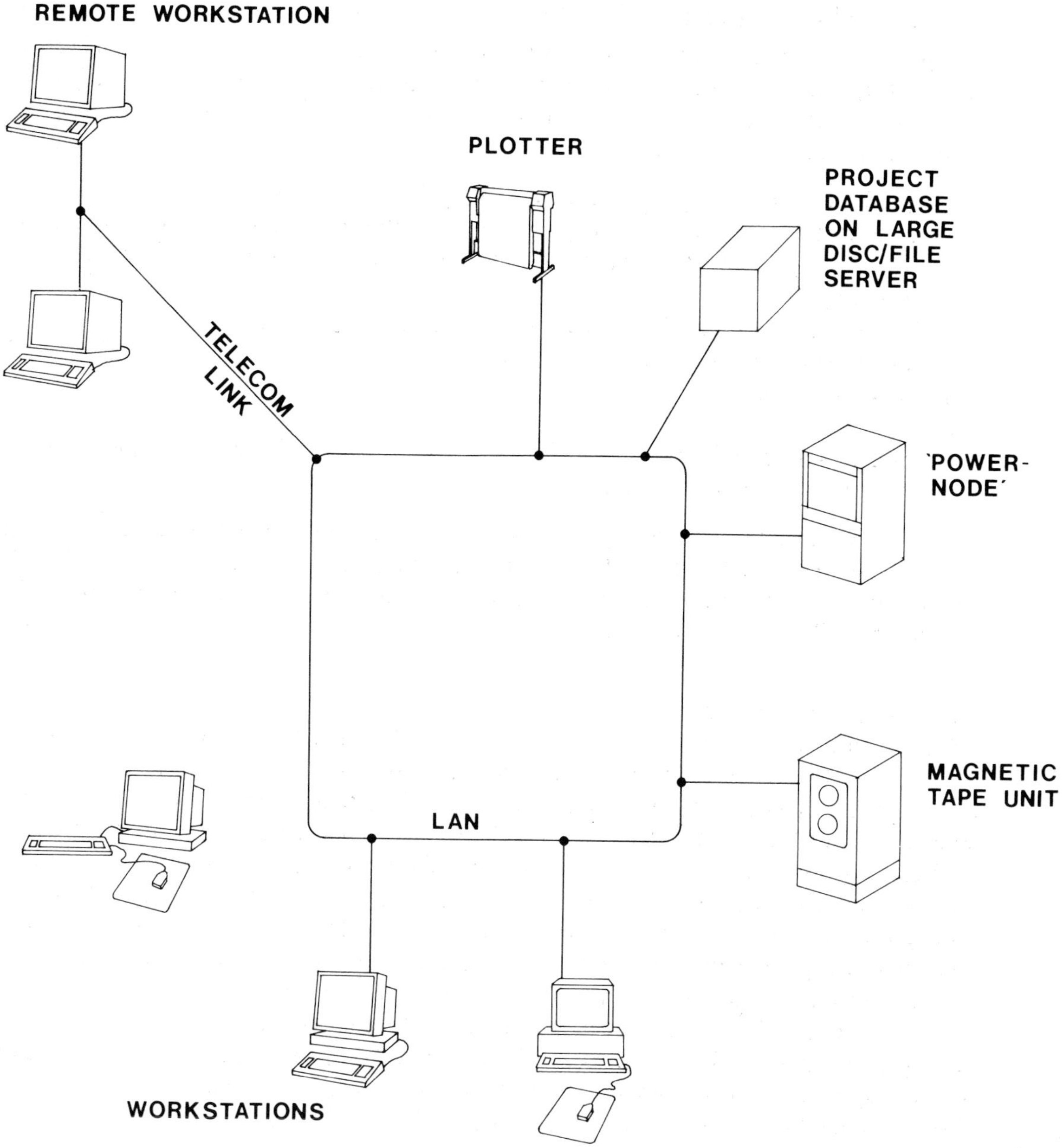

REMOTE WORKSTATION

PLOTTER

PROJECT
DATABASE
ON LARGE
DISC/FILE
SERVER

TELECOM
LINK

'POWER-
NODE'

LAN

MAGNETIC
TAPE UNIT

WORKSTATIONS

Fig 6.1 A distributed system.

Management must ensure that individual operators do not create their own files of design data independently on their own nodes, for that is the road to anarchy.

In hardware terms, flexibility is enhanced because such a network can at the outset be a small low-cost entry system. Later it can be rearranged or extended in a variety of ways to match the new or growing needs of an organisation. The merit of this arrangement is its flexibility to meet an always-uncertain future. It seems likely to be the predominant hardware solution for CAD and other applications in the future.

Distribution of hardware and tasks depends on the ability to communicate data. There may be no difficulties when the equipment is sourced from the same computer manufacturer. When a variety of equipment is to be networked, data communications depend on communications standards and compatibility. Progress in the imposition of such standards has been slow and so it must not be assumed that any hardware and software items can be connected up into a workable network. It is essential to choose only the network products having an established track-record.

Distributed systems need not be seen as a new threat to existing computer systems installed within a company. There may be no need to scrap everything so that the new distributed hardware approach can be implemented. The concept of *Heterogeneous Networks* can in many cases come to the aid. Here the company retains its existing equipment, and seeks to upgrade it when necessary into a distributed system. It does this by adding workstations, extra computers or peripherals together with a suitable communications network.

6.4 ACCOMMODATING THE COMPUTER AND PLOTTER

Large installations

Special accommodation for the computer is something that has to be considered only for large computers like the supermini or mainframe. Then all the paraphernalia may be needed, such as:

- enclosed computer room,
- false flooring and/or trunking for cables,
- air-conditioning for the fine control of temperature, humidity, and to filter dust and smoke particles in the air,
- smoothed electricity supply with plenty of outlets,
- smoke and heat detection; controls and alarms; fire fighting equipment suitable for electrical installations,
- racks or storage space for consumable materials, such as printer and plotter paper, pens, ink, magnetic tapes, and for important manuals,
- controlled access to computer room.

In the biggest installations, some back-up facilities in the form of duplicated or stand-by equipment might be needed. These can provide resilience to cope with problems and failures. The control console for the computer would be placed in the computer room too, but the user-workstations should never be accommodated there.

Large windows should be let into the partitions of the computer room so that staff can see what is happening there without having to enter.

Armer[16] advised that power lines to the computer should originate from a main power distribution panel and be laid in metal conduit dedicated to this purpose. For a large machine, it may be necessary to install an electrical noise filter or constant-voltage transformer in the supply circuit. Advice should be taken on the installation of data cabling. It is prudent to keep data cables separate from power cables. Allowance must be made at the time of installation for future expansion of the computer system. Flush fitting data sockets with spring lids fitted into the floor can be an appropriate arrangement in a large office. Armer also advised that cables routed outside a building must incorporate protection against voltage surges caused by lightning with barrier devices employed to protect the transmitter and receiver[16]. The services of a data cabling contractor can be valuable as he may well have more relevant knowledge than the normal electrical contractor.

General information is available on computer accommodation[17] while the supplier will be able to provide assistance in drawing up detailed plans. Obviously the accommodation for equipment must be made available, and checked with the supplier, before the equipment is shipped.

Office environment systems

Smaller minicomputer systems, workstations and PCs can operate within wide environmental limits. They are often described by their suppliers as 'Office Environment' machines. This does not necessarily mean that they can be placed within the normal office, without first carrying out a few checks.

The power rating of all the equipment should be added up to give some idea of the extra waste heat that will output into the office. In a large space this might dissipate satisfactorily. In a small room, it could make the area oppressively hot in summer or overwhelm an installed air-conditioning system. This might not be detrimental to the computer equipment, but may be very unpleasant for people working in the same room.

These machines can usually operate from the normal power outlets in the office.

Some of the office-environment minis produce quite a lot of noise from their cooling fans. This might bother workers in the same room.

Plotter accommodation

Pen-plotters function by automatically drawing each line or text character, one after the other, until the whole drawing sheet is completed. This process takes perhaps thirty minutes or more, depending on the plotter performance characteristics, the size and density of the drawing. The plotter draws to an accuracy of 0.1 mm or better. Plotter paper will expand or contract if changes occur in the temperature or humidity during the time of plotting. Taking all these facts into account, slight changes of scale can occur within the drawing sheet. These may or may not be noticeable or significant. When lines do not join up with lines drawn much earlier, or when there is crossover where a sharp corner is expected, the effect will be immediately noticeable. We shall see in Section 6.9, where we discuss plotter consumables, that a stable medium can be used in place of organic paper, when necessary.

The normal office environment may prove to be too variable for a pen-plotter, particularly when nearby doors and windows are liable to be opened and closed. It may then be better to place the plotter in its own room and to keep the door closed whenever possible. Again, a glazed-partition permits the plotter's progress to be monitored from outside. A pen-plotter creates some noise, and this can be another good reason for banishing it to a separate room. In practice the plotter and computer can be accommodated together.

It is advisable to store unused rolls of plotter paper in the same room. Then they can become stabilised to the prevailing temperature and humidity. Indeed new rolls should be unpackaged and loosened within the controlled environment to encourage this stabilisation process. Jenkins[18] recommends that the ideal conditions are 18–24° Celsius, at a relative humidity of 45–55%.

If the atmosphere in the plotter room is too dry, then trouble may be experienced when plotting with liquid ink using a fine nib. A clotting pen can of course completely ruin a plot. Some missing lines might go unnoticed and this is potentially very serious. If such problems occur, a cure may be to install a humidifier in the plotter room.

The t^2 systems have a facility which can limit the effect of environmental changes on plotting. This enables a large drawing sheet to be plotted in strips, thus allowing the paper ahead to settle somewhat before it is plotted on.

Electrostatic-plotters produce drawings very quickly and so are found in systems where there will be a large volume of drawing output. They are quiet in operation and require rather less support and attention compared with pen-plotters. Electrostatics also benefit from a clean and fairly stable environment.

6.5 SITING OF WORKSTATIONS

The workstations can be either concentrated or dispersed within the organisation, depending on individual preferences. There seems to be no overall consensus on which course is the better. The following points may help with planning.

Concentrated workstations

Arguments for grouping all the workstations in one special location are as follows:

- If all the operators report to the CAD co-ordinator, he is able to keep better management control over their work. This is perhaps the key issue.
- An office area can be specially fitted out to provide good ergonomic and environmental conditions. Comfortable working conditions will have beneficial effects on work output, remembering that CAD operation is very intensive work. This is a subject to which I shall return in more detail in the next section.
- All operators could be located near to the plotter.
- It can be easier to cope with peaks and troughs in the organisation's workload, because resources can be rapidly switched to the urgent projects.
- Users are taken out of their normal work environment and this is an encouragement for them to prepare well for each work-session. Operators will be less subject to interruptions, especially if telephones are not installed in the workstation area.
- Better conditions are available for training sessions. With two stations located close to one another, a new operator can gain experience while remaining within range of an experienced colleague.
- All operators benefit from being able to refer to each other when problems or new situations arise.
- Public Relations. This is where the visitors are taken. Wall space could be provided for the exhibition of examples of CAD output.

These seem to be a powerful set of arguments for concentration. Concentration can provide a much-needed sense of identity for the group. There is a danger however that it will be hived off in an obscure corner of the organisation, or to a separate building. Then the group would become isolated from the remainder of the company and 'forgotten'.

Dispersed workstations

The workstations could be dispersed around the design offices with each located close to the appropriate project design team. There are some arguments in favour of decentralisation:

- CAD will be more rapidly accepted and integrated into the organisation's mode of working. This is because everyone will appreciate better what is going on. We must remember that it is not just the operators that 'use' CAD. The project leaders will be able to understand and control their work better. Technicians operators can work near their designers.
- Operators do not have to move all their reference material a long way to the workstation at the start of each session. When they lack any information, the other project people are not far away.
- Flexibility. Workstations can be moved around fairly easily to where the workload lies, provided of course the cabling arrangements allow this.

- Where the workstations users report to project leaders rather than to the single CAD co-ordinator, this dispersed arrangement can be preferable. Where they all report to a single CAD co-ordinator, this arrangement may lead to poor control.

6.6 PLANNING THE OPERATING ENVIRONMENT

A trained CAD user and a workstation together form a high-cost combination. A work-session at the terminal should be a period of intensive work. Management will rightly demand much from the operators. It should appreciate that it is in everyone's interest that a good operating environment is provided.

This is certainly not an academic point or even merely a matter of aesthetics. During each work-session there can be a significant fall in operating efficiency and an increase in the incidence of errors. The cause is fatigue setting in after a period of intense concentration. Careful attention by the computer manager to equipment layout, to accommodation, the furniture, fittings, lighting and other environmental conditions can help considerably. The cost of some of the measures now to be discussed is not likely to be prohibitive, especially if they are undertaken at the outset. The return on this investment will be in the form of increased operator output, and it is likely to prove attractive. Many of the following points are taken from an article[19] by Port and Tamplin which deals with this subject.

Orientation of the room and lighting

A relatively low level of general lighting is best and screen reflections must be avoided at all costs. It follows that direct glare arising either from sunlight or from artificial lighting must be avoided.

When it can be arranged easily, a north-facing room is best, because then the natural lighting is indirect. A south-facing room does not suffer too badly from direct sunlight, except near the window, but conditions can become oppressive because of solar heat gain.

East- or west-facing rooms suffer most from direct sunlight, especially in the morning or late afternoon respectively. The best defence is to install blinds with adjustable slats or window shades which are able to diffuse the direct light. Black-out curtains that do not filter the light create conditions of high lighting contrast and are not so suitable.

To avoid getting reflections off the computer screens, they should be placed at right-angles to the window. Alternatively the workstation may be arranged so that the operator sits facing the window. In this case the lighting contrasts are reduced when the equipment is placed deeper into the room.

Troublesome solar temperature gains may have to be countered by external shades, air conditioning or perhaps by quiet fans or extractors.

Artificial lighting must be glare-free. One solution is to provide a ceiling light with individual dimmer control. This should be above the workstation and not behind the user's chair. This light should be small in source area. Harsh fluorescent strip-lights must be equipped with diffusing covers, although even with these they generally cause distracting screen reflections. Uplighters which bounce light off a pale-coloured ceiling can provide a better solution. Where several workstations share the same room, each can be provided with an angle-poise lamp fitted with a small spotlight bulb. Then each operator can adjust the illumination level on reference documents without affecting anyone else.

Seating

Too often, the importance of the operator's seating is overlooked. The very nature of CAD means that draftspersons no longer have regular changes of body posture. There is no need for actions that once interrupted the flow of work. No longer does the drawing board need adjustment, paper need changing, pencils need sharpening, plan chests need visiting. Such actions helped to alleviate the problems induced by bad posture at the drawing board. With CAD, the degree of mental concentration is increased with minimal breaks or shifts in position. Any CAD operator experiencing discomfort, aches and pains during the span of a normal day should first check the suitability of the posture adopted before looking elsewhere for causes.

Ergonomically the chair is a very important element of the workstation. Investment in a good chair brings returns in increased operating efficiency.

The main feature of the chair should be a high contoured back, curved to fit the shape of the spine. It should be adjustable relative to the seat. An adjustment of just a few millimetres here can make all the difference between longterm comfort and the gradual onset of back pains, neck and shoulder aches and even eye strain. The high back supports the shoulder blades and this contact enables easy turning of the chair towards adjacent work surfaces.

The chair should have gas lift adjustment, variable rake for the backrest and a five-star base for stability. Castors or wheels are best on carpet, but can be dangerous on a hard floor surface where slides are preferable. Arms on the chair are a mixed blessing. They are useful for support and swivelling, but knock into table edges or prevent the user sitting close enough to the workstation.

Height adjustment of the chair is the most important feature. The operator's eye level should be such that he looks down at the centre of the graphics screen at an angle preferably of about 15 to 30 degrees below the horizontal. The height of the operator should also be such that with hands on the detachable keyboard or data tablet, the upper arm and forearm should subtend a right angle. Keyboards are normally placed at 675 mm above the floor.

If the same workstation is used by different persons, then the computer manager must encourage each person to make all the necessary readjustments to achieve the right ergonomic conditions. Operators must not carry on using the equipment as vacated by the last person.

Furniture and fittings

The screen should have both rotational and tilt adjustment. It should be regularly cleaned with the recommended solvent. Workstation on/off switches should be easily accessible.

Positioning of workstations must allow the operators ease of access into the 'driving seat'. Once there, they should be free from being crowded by passing traffic. The conventional drawing office often caters for this by the placing of drawing boards as cul-de-sacs leading off a main access route. A similar arrangement for workstations is appropriate. Partitions or acoustic screens should be fitted tightly together with no gaps.

The following items should be provided in the workstation room:

- plenty of work-tops for layout of reference documents;
- shelving for manuals (this helps to prevent the manuals from 'wandering', which is always annoying);
- fire extinguishers suitable for electrical risks;
- wall clock;

- spare office chairs for visitors, trainees, etc;
- enough coat hangers for operators and visitors;
- at least one large noticeboard with a dull colour matt finish. Care is necessary, because white paper notices attached can be a cause of screen reflections.

The walls of the room should be painted with colours of a dull matt finish. In practice, much wall space is likely to be used for display of notices and examples of work. Non-static carpets or mats are also desirable particularly if a dry atmosphere is a problem.

If the operators do not have their base within this room, then telephones should not be installed. Operators must not be subjected to interruptions by outsiders when using such expensive facilities.

Fig. 6.2 illustrates a suggested layout of a room to contain two workstations with additional facilities for training and demonstrations.

Power and communications wiring

The installation of power and data cables within the fabric of the building has been discussed briefly in Section 6.4, and more detailed advice is available from Armer[16]. Let us now turn to the office accommodation and workstation area.

Even when the workstation and attached equipment can operate from a single standard power outlet in the office, provision should be made for five outlets per workstation. Ancillary equipment might well include a hard copy unit, electric fans, angle-poise lamps, test equipment for hardware maintenance, and cleaners' equipment. Ensure that each item is correctly fused. Workstations may require special measures to protect them from power surges.

Power outlets and communications ports must be accessible with ease, preferably without stooping. They should not be obstructed by other equipment or furniture.

Power and data cables should be long enough to allow adjustments to be made in the positioning of workstations, without any wire being at full stretch. Spare cable should be neatly secured behind the display screen or attached to walls. Systems furniture often has wire management channels incorporated, so wires can be placed safely out of sight without becoming dust traps, yet they are always accessible when necessary. Cables preferably should not lie on the floor, and certainly should not lie across access routes where people can trip over them. If this is *absolutely* unavoidable, suitable covering or channelling must be arranged.

Communications ports or *patch boxes* and the related cables, plugs or connectors should all be clearly marked with information such as workstation designation or line numbers of the host-computer. It is the responsibility of the computer manager to organise this and to make sure that all cables remain tidy.

6.7 STARTING, STOPPING AND OPERATING THE HARDWARE

These points are never quite so trivial as might be supposed. There are several different items of equipment, and each has its own procedures for switching on and off. Paper in the printer or plotter might have to be changed, or the plotter pens might run out of ink. Then there will be a procedure for starting the CAD software running. Each operator may have to log in and log out at the beginning and end of each work-session.

The computer manager must ensure that each operator knows how to deal with, at the very least, the commonly-occurring procedures and simple fault conditions.

The computer manager should devise sets of instructions and place these near each item of equipment – making sure also that they remain available there at all times. These

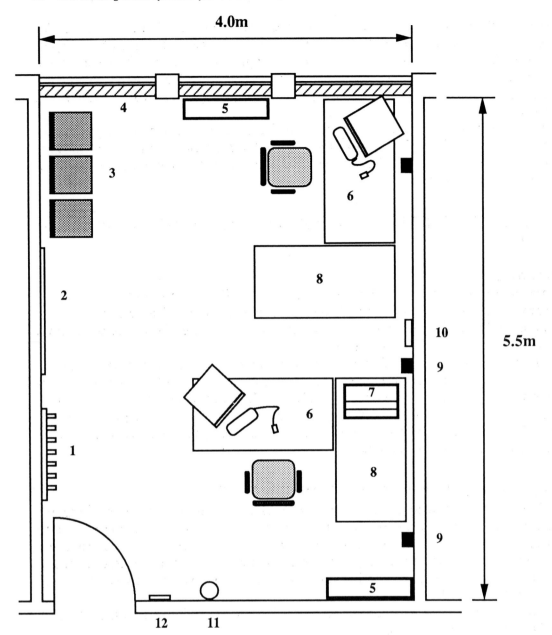

Suggested CAD operating environment with provision for training and demonstration

Legend

1 Coat hooks
2 Notice board
3 Spare seating
4 Window blinds
5 Bookcase
6 Workstation

7 Hard copy unit
8 Reference table
9 Power points
10 Wall clock
11 Fire extinguisher
12 Lighting controls

Fig 6.2 Suggested layout of a CAD workstation room.

instructions should be in the form of checklists. They can be printed on a particular colour of paper, so as to be instantly identified, and protected in clear plastic envelopes. However, instructions in words are not always enough, and there is merit also in holding occasional practice runs. All this is necessary because problems have a habit of occurring late in the evening or at other times when, for example, an operator is alone and trying to get vital work finished.

The computer manager is also responsible for ordering and maintaining stocks of hardware consumables. Examples of these include printer ribbons; paper for printers and plotter; pens and ink for the plotter; toner for electrostatic plotters; magnetic tapes; extension cables for data and power.

6.8 DATA SECURITY, DISC BACK-UPS AND ARCHIVING PROJECT DATA

Passwords

Password systems are built in to many CAD systems. They have a positive purpose, of ensuring that access to project data is only available to authorised operators. Similarly there are other high-level programs or functions which should be restricted to those people who have a system management function.

All this is reasonable. The design data which is built up within the computer takes on the value equivalent to the cost of the operators and machine-time involved in its creation. Clearly this value can be very high indeed, and the data must not be placed in jeopardy. Responsibility for instituting and undertaking the necessary procedures lies with the computer manager. If any operations are delegated, a close rein must be maintained to ensure that all is well.

Disc back-ups

Project data and programs are stored semi-permanently on the disc store of the computer system. There are many types of accident which can cause the information on disc to become lost or corrupted. There may be a failure of the disc itself; an operator may make some serious error, or a whole project or several components may be deleted by mistake. Taking due account of the high value of this data, it is imperative that copies are made of it at regular intervals.

Disc back-up is the procedure for routinely making copies of the entire disc contents to avoid complete loss of valuable data.

Such back-ups are performed according to an agreed cycle. The computer manager must set up a suitable procedure, possibly with advice from the CAD supplier. For example, the entire contents of the disc might be copied to magnetic tapes every other working day. Where there are no magnetic tapes, back-up would be to exchangeable disc packs, or to several floppy discs (this is a slow and time-consuming process).

An alternative type of back-up cycle is possible on some computer operating systems. Here the entire disc contents are copied less frequently, say once a week or even every two weeks. In between these times, say every day, the computer can make a copy of just those computer files which have been changed in some way, i.e. they have been active, since the last full copy was made. This slightly more complex procedure can be performed much faster than simple back-ups. However on the (hopefully rare) occasions when it is needed, disc recovery would be a little more involved.

This is risk management. The frequency of disc back-ups must balance the risk and consequences of an accident, with the time and trouble involved in performing the back-ups. If the contents of a disc are lost, all work carried out since the last back-up was done will be lost and so all this work will have to be repeated. Normally back-ups are done in the early morning or in the evening when no other users are active. This is because back-up tends to slow system response.

Sometimes three sets of magnetic tapes are maintained and used in rotation – the new data is always written on top of the oldest copy. This is the *grandfather-father-son* scheme. Some installations keep many sets, perhaps not overwriting data for a full month.

Procedures for restoring the contents of the disc should be practised occasionally.

As computer equipment and computer operating procedures become more reliable, a feeling can often set in that too many 'useless' back-ups are being done. Sometimes there is a tendency for the cycle to become lax and back-ups are missed. It is the responsibility of the computer manager to see that this never happens. Anyone who has suffered the effects of a disc failure understands why this is so important.

The back-up copies must be stored either in a separate building from the one which houses the computer, or in an (effective) fire-proof cabinet. The computer manager must maintain a log of the dates and times when the security copies are made, together with the tape or disc serial numbers.

Archiving project data

Archiving is the name given to the process of making copies of specific project files for long-term storage. Normally the copying is to magnetic tapes or floppy discs.

The reasons for archiving may be project-related. At critical stages in the design, there may be a wish to retain a permanent record of all the data current at that time so that it is accessible if required later on. Obviously when the project is complete, another copy would be made at that time. When a project has merely become dormant for a while, the computer manager might well decide that he needs the disc space which it occupies for other purposes. Then the project may be archived and afterwards can be deleted from disc. At any future time, of course, it can be restored from an archived copy.

Archiving of inactive project data should be considered when the disc becomes more than 80% full.

It is essential that at least two copies of important data are maintained in separate locations, where, for example, a fire cannot destroy both copies. Data that will be retained on the computer's disc could be copied once, and the copy stored in a fire-proof cabinet or in a separate building. Data that is to be deleted from the computer's disc must have two copies taken. One can be kept in the computer room, the other in a separate building. The maintenance of two copies of everything provides some measure of protection against operator errors.

A log must be kept of archived project data. Each entry must contain model or drawing names; date; tape serial numbers; name of person making the archive. If careful logs are not kept, there will be difficulty in finding the data again.

A clearly defined procedure for restoring archived information to the disc store must be established.

Data stored on magnetic media is not completely stable over many years. The computer manager should see that tapes or discs are recopied occasionally, preferably to a set pattern. The advice of the supplier should be sought regarding this matter.

Disaster contingency plan

Every computer manager must work out and document his own 'Disaster Contingency Plan'. Moreover, the CAD director should ensure that such a plan exists.

To produce such a plan, the computer manager must try to think of everything that could go wrong. For each of such events he must decide on, and document, the most appropriate course of action. In most circumstances it would be relevant to list the symptoms of the fault plus the name and telephone number of the person or organisation who can make the necessary repairs.

Part of the disaster contingency planning might involve making certain prior arrangements when there is another organisation in the vicinity which is operating the same type of CAD system. For example, in the event of serious plotter breakdown, some plot files could be transferred on a magnetic tape. Then some urgently-needed drawings could be plotted perhaps with only a minimal delay. It is in the best interests of the computer managers of different firms, even those that may be fierce competitors in other respects, to co-operate with one another in this way.

The sources of all the types of spares and consumables should be noted. If there are friendly firms where such consumables can be borrowed in an emergency, this too is worth recording.

The system supplier can be of assistance when drawing up such a plan. Several people within the organisation – in addition to the computer manager – should be aware of its existence so that they can act appropriately. A little time and prior thought can pay handsome dividends when the emergencies eventually occur.

6.9 PLOTTER MANAGEMENT

Despite the fact that a plotter is a precision instrument, it is often the least satisfactory part of the hardware. Careful attention to detail helps to get the best out of it. Sometimes one operator or even an office junior is trained to look after it, but of course it remains the responsibility of the computer manager.

The plotter can make heavy demands on the computer processor and it is often worth forming a queue of plots to be run off either in the early morning, in the evening or even at night.

Some of the following information on plotter management is derived from articles by Jenkins[18, 20] and Tope[21].

Pen plotters: paper, pens and ink

Environmental conditions for plotting have been discussed already in Section 6.4. Successful plotting also depends on using suitably matched combinations of paper, pens and ink.

Organic papers are susceptible to changes in temperature and humidity. Types used are:

- Detail paper: Cheap and used commonly for preliminary design and for one-off drawings.
- Cartridge paper: Expensive, has better contrast and is used for presentation work.
- Tracing paper: Rather unstable; used where prints will be made.
- Translucent paper: Prints with a mottled image. For dye-line or diazo prints.

Synthetic media is impervious to changes in humidity. It is strong. These properties make it preferable for archived drawings or for precision work. Synthetic media must be handled with care, preferably with gloves, because ink may not adhere to regions where there are fingerprints. Types used are:

- Polypropylene film: Stable, strong. Popular and not too expensive. Translucent media is used for dye-line and diazo prints.
- Polyester film: Stable, strong. The best but very expensive. Available in matt or clear form.

Roll-feed plotters take paper rolls which are normally fifty metres in length. Requests can be made for many plots in one session, and so the plotter can work continuously for a long spell – perhaps throughout the night. However, problems may be experienced with long plot runs using liquid ink, and large capacity ink reservoirs may be necessary.

Sheet-feed plotters have to be fed with a single sheet each time, so long unattended runs are not possible. For this type of plotter, drawing blanks can be pre-printed with a frame and the company's title block. This saves some plotting on each sheet.

Types of pens are:

- Liquid ink pen points: These are refillable pens. Nibs are normally chromium or stainless steel, but tungsten tipped pens are expensive but last longer and are preferable on polyester (which is slightly abrasive). These can draw in various line widths: 0.2 mm, 0.3 mm, 0.4 mm and 0.6 mm. These pens are not very good on detail paper, or for long unattended runs. The thinnest nibs are unreliable. Better results are possible if the plotter speed is restricted. Pens must be kept clean.
- Ball points: These produce a reasonable, consistent line, but being oil-based ink, it does not dry and so is liable to smudges. Only a single line thickness is available. Being rather thin, this can be a little weak when printed. There is some tendency to skid on film. Available in a few colours. Long lasting, convenient to use and cheap.
- Pentels (roller ball):   Water-based ink so there is little smudging. Line thickness can be varied somewhat. 0.2 mm is often used. Even the extra fine points give a good strong line, especially on detail and translucent paper. There is a tendency to skip on film. Available in a few colours. Fairly long lasting and cheap.
- Fibre tips: These give excellent line intensity on a variety of papers. Lines are rather broad, and the tips broaden with use and they wear out fairly rapidly. Fine detail may be lost. Not very suitable for a lot of annotation work. Available in many colours and suitable for area-fill and rendering work.

Liquid ink is best for the highest quality or final drawings. It is rapid drying, uniform and opaque. The viscosity necessary in ink limits the usable drawing speed of the plotter. Ink for organic paper is usually water-based whereas that for synthetic media may be solvent-based. Inks based on chemical dyes are less viscous, so the plotter may operate faster, but they produce less contrast than Indian ink. Plotter ink is available in black and about six colours. The reproducibility of the colours when printed using dye-line or other reproduction methods varies and should be checked. For example, red or brown ink reproduces as grey. In practice, this can be useful for some purposes to contrast with the normal black – for example for grid lines or for showing existing construction.

Electrostatic plotters: paper and toner

The electrostatic process operates by the placing of electrical charges at the relevant points on the coated surface of the paper. This is done as it passes over a writing head in the plotter. Toner liquid is applied and minute grains of black resin and carbon adhere to the charged points. Colour electrostatic plotters are available. The same process is used but there are four toners applied at different times – black, cyan, magenta and yellow – to produce the whole range of colours.

Special dielectric-coated paper is needed to allow the electrostatic process to work. Report grade paper is white and opaque. Good contrast copies can be produced with it for photographic or microfilm reproduction. Translucent vellum is used for dye-line or diazo copying. For archiving and precision work, there are coated polypropylene and polyester films available.

Toner must be renewed at intervals depending on the volume of use, and as recommended by the plotter supplier. Jenkins [18] recommended that at least once per day, the writing head, backplate and rollers should be wiped clean. Once a week, the toner circulatory system should be flushed round for ten minutes with clear dispersant liquid.

6.10 DATA EXCHANGE WITH OTHER ORGANISATIONS

Data exchanges can be divided into three classes:

(1) Exchanges between organisations that each run the same CAD system.
(2) Exchanges between organisations using dissimilar programs. For example, a survey firm using its own survey programs may wish to send site data to an architect who then has to enter this into a CAD system.
(3) Exchanges between organisations running dissimilar CAD systems, using either a conversion program or IGES.

Let us look at each of these in turn.

(1) Exchanges between the same type of CAD system

These will present the fewest technical problems but there are significant issues to be tackled. In particular the respective users working in parallel on a project ought to co-ordinate their use of the data classification systems.

Where the two organisations are in practice using different classifications, it is possible – but with much inconvenience – for the recipient to change the data into his own classification. He still needs to know the details of the sender's conventions. Pre-planning on the project can aleviate these problems.

Since the same CAD system is in use in both places, the data will be in an acceptable format. Data is normally physically saved on a magnetic tape by the first computer so that this can be read by the second computer. If the computers are different models, the tapes would have to be written and readable in industry-standard format. Transferable discs or direct transfer by means of a telecommunication link might be other options, depending on the two hardware configurations.

(2) Exchanges between organisations using dissimilar programs

In this case, the source program may have been arranged so that it can write its output data in a format which suits the destination program. This of course not very likely.

In other circumstances, the source program must output its data to a computer file which we shall call the *Source File*. A computer *file* is merely a batch of data which can be stored or transferred within the computer as a single unit.

Next, a conversion program, generally known as an *Interface Program*, must be available. This must read the data from the Source File and convert it into the format that will be acceptable to the destination program. It writes it to a file which we can call the *Destination File*. The interface program will run on the source or destination. The exchange in one direction is shown in Fig. 6.3.

The developer of the Interface Program has to know the formats of the data in both the source and destination files.

Fig 6.3 Data exchange via an interface program.

(3) **Exchanges between different CAD systems**

With the proliferation of CAD systems, it is becoming increasingly common for one design office to want to transfer design data to another office where an incompatible CAD system is in use. These transfers often prove to be difficult. The problem lies in the fact that each CAD system structures and stores its data in a different manner. Because of the difficulty, a bottleneck will form in CAD data communications between firms.

Sometimes a direct conversion or interface program exists, specifically developed for the transfer of data from System A to System B. In this event, use of this direct interface is usually the most efficient means of transfer, and this is what we shall look at first.

A prior-agreement between the two companies on their classification systems is to be strongly encouraged. Nevertheless in practice this will be far from easy. The two CAD systems will have been developed independently of each other, each having its own terminology. For example one system might have layers and categories, while the second only has components or shapes. There are certainly going to be some difficulties when data is to be transferred out of a 2–D system into a 3–D system, or vice versa.

Back-transfers from System B to System A would need a different interface program. It is important to note that any such interface used must relate to the particular versions of the two CAD systems between which data has to be transferred.

Provision of such direct interface programs for data transfers may seem to be the most logical solution to a difficult problem. It may be appreciated however that with many incompatible CAD systems now being used, the number of possible transfer paths increases rapidly. So a very large number of direct translation programs are potentially needed. For n systems there would have to be $n(n-1)$ translators. As the CAD systems are upgraded and improved over the course of time, different versions of these interface programs would have to be maintained. The problem soon becomes unmanageable in practice, and so IGES was introduced to provide a better solution.

IGES is the Initial Graphics Exchange Specification. This was first developed in the US aerospace industry in 1978. It documents the ways in which CAD primitive elements like points, lines, arcs and text characters are to be stored within a standard intermediate or 'neutral' file. This is a file which can be set up between two CAD systems.

To permit IGES transfers, each CAD vendor is required to write two IGES translator programs. One must be able to convert any graphical data from its own system into the IGES neutral format – this is a *pre-processor* program. The second is an IGES *post-processor* program which must be able to translate any data from the neutral format into the format of its own system. So for *n* systems there would have to be 2*n* processors. The vendor is responsible for modifying its own pre- and post-processors when it makes any changes to the data structure of its own CAD system.

An IGES data exchange can only be effected if both system vendors support IGES. First the pre-processor program belonging to the source CAD system is run on the source computer. This produces the IGES neutral file. This file is transmitted, for example on magnetic tape or by telecommunications, to the destination computer. Here the post-processor belonging to the destination CAD system converts the IGES neutral data into a format suitable for this system. The forward and back conversions between two CAD systems are shown in Fig. 6.4.

IGES transfers work tolerably well when 2–D drawings have to be exchanged. Nevertheless IGES has been plagued with problems. The basic problem is the mismatch between graphic entities in CAD systems which are structurally very different. Also, IGES is a specification written in English and so is always open to interpretation. Particular graphics entities may or may not be supported in particular pre- or post-processors.

It is vital that the sender and receiver should establish prior agreements on classification conventions, line styles, symbols and so on, but still there are many opportunities for unexpected problems to occur. The neutral drawing files tend to become very large indeed, and much computer time and resources are needed to effect IGES transfers.

Before any communication work is attempted via IGES or interface programs, the respective computer managers should undertake a test exchange between the two systems. A rigorous check is to transmit a drawing in one direction, and then transfer it back again. The end-result should of course be identical to the original.

From this discussion, it may be seen that data exchanges are possible if the respective vendors have the appropriate pre- and post-processors. Exchanges are being made but they usually take some time and trouble. The situation is likely to improve gradually as more experience with IGES is built up.

IGES 3.0 and later versions have been specified for 3–D transfers but eventually IGES is likely to be replaced by standards like PDES (Product Definition Exchange Standard) or STEP (Standard for the Exchange of Product Data).

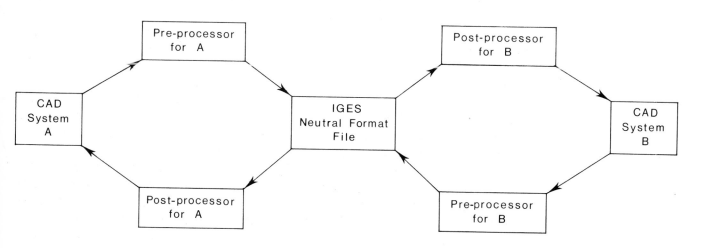

Fig 6.4 Data exchange via IGES.

6.11 SYSTEM MAINTENANCE

Normally a maintenance contract is agreed on an annual basis between the customer and the system vendor. This covers:

- preventive maintenance of the hardware,
- repair of hardware failures,
- correction of software errors,
- support to the customer when any acute difficulties are encountered in using the system.

In most circumstances, the supplier will want to deal with a named individual in the user firm. This would normally be the computer manager.

When the computer can be linked to the public telephone network, this can be useful for system maintenance. Diagnostic tests can be run to locate hardware or software faults, without the support people even having to journey to the computer. Even if a visit becomes necessary, they are more likely to arrive with the correct replacement components.

Preventive maintenance of hardware

Preventive maintence of the various items of hardware involves a pre-defined number of routine visits by a technician. The dates for these visits should be agreed well in advance and entered into a system diary. The dates and times of actual visits should be recorded too, together with any actions taken and the hardware items dealt with. Any missed visits must be chased by the computer manager.

Clearly some disruption to production work with the system can be caused by these visits, but this should be minimised if forward planning is being done. This preventive maintenance is of course very important if the risk of serious breakdowns is to be reduced.

Hardware failures

When breakdowns do occur, the cause must be isolated as far as possible, and the service organisation called by telephone. Usually the maintenance contract stipulates the maximum interval within which the service organisation must respond to a call. A suitable fault logging procedure should be set up by the computer manager. A record should be made in the log or in the system diary of:

- name of person reporting the fault,
- the nature of the fault,
- date and time of call,
- date and time of response,
- date and time of repair completion,
- the nature of the repair made.

These records can prove invaluable in the future if similar faults should recur.

Software faults

Software faults, actual or suspected, should be noted on a separate recording system. These records can include:

- Name of operator reporting.
- Date and time of fault occurrence.

- The operator should log with some detail the circumstances of the fault. This includes what operation he was undertaking immediately prior to the problem, the exact effect of the fault, and perhaps the action he has taken to overcome or recover from it.
- Software problems can be classified in one of two categories – *Chronic* and *Ordinary*.
- Date and time of reporting to the software supplier.
- Date and time of the software correction.
- Nature of the software correction.

Chronic faults are deemed so serious that it is necessary to telephone the supplier in the expectation that a solution can be effected as soon as possible.

Ordinary faults are less serious, and are either not very likely to recur, or they can be easily avoided in future with some simple 'work-around'. The logs of ordinary faults can be collected and transmitted to the supplier at set intervals by the computer manager.

The records prove invaluable in tracing seemingly intractible problems. There is a tendency for the same software faults to recur, and the records may give details of suitable evasive actions which have been successful previously.

Support of users

There are occasions when a user gets into difficulties and needs advice from the supplier's experts. The supplier is entitled to assume that the users have been properly trained and that efforts have been made to find solutions in reference manuals. Requests for help should be made through the proper channels – normally the computer manager. These requests too should be logged by the computer manager, because these problems also have a tendency to recur. Previous logged notes might solve a problem without reference to the supplier.

6.12 LOGGING OF COMPUTER RESOURCES USED

Records should be kept of the system resources that are used. Such records might include, for example:

- date, and workstation session start and finish times,
- operator's name,
- project worked on,
- nature of work, e.g. component creation, model assembly,
- sheets plotted,
- large print runs done e.g. schedules.

Clearly such records are useful. Analyses carried out by the computer manager over a period provides information which helps in the control and management of the CAD system itself.

The information will also be useful for project management. It could form part of the internal cost system where manpower and computer resources are costed to individual projects. It might contribute to the external fee calculation when the computer resources can be charged directly to a client.

Some minicomputer systems have in-built log-in and log-out procedures to record work-sessions. These can provide records, for example, of processor time and disc access time used, and the amount of disc storage in use during each work session. With suitable rates applied, a total cost can be calculated for each session, and assigned to the relevant jobs.

Unless programs exist to cope with this entire costing process, this last method is probably over-elaborate. Internal computer resources are not well understood by many users, and this method often leads to arguments where abortive work is involved.

It is usually better to use a very simple system. This can be done by applying suitable average charge rates against a few recognisable items like workstation-hours used and the number of sheets plotted.

In practice, costing strategies can strongly affect in unforeseen ways the sentiments for CAD among staff. Working with some pre-set rates, a system might appear to be very expensive to use. This is worse when operators are learning or gaining initial experience. Project managers can become worried. Many staff even fail to understand the subtle difference between internal costing and external accounts. For example, use of the system might be banned by one project manager because the costing formulae make it appear rather expensive. In practice the system (once purchased) costs the organisation almost as much to leave it idle as to make full use of it. Non-use might even mean that additional fees that could have been claimed are in fact foregone.

Sometimes special costing strategies are set to try to overcome such problems. For example the system might be provided free-of-charge to projects during the first year only, when efficiency of use might be rather variable. The system is costed to the firm's overheads during this time. It is not prudent to allow this 'cost-holiday' to continue too long however, because resources that are not clearly costed tend in the long run to be wasted.

Another strategy is to disperse workstations to departments, and then charge each department a lump sum annually for each station. This has the merit of being simple, and it encourages efficient use.

6.13 SYSTEM UPGRADES

The importance of choosing a open and flexible system was stressed in Section 4.7. Flexibility in this sense relates to a system that can be enhanced or upgraded in a variety of ways to suit an always uncertain future. The merits of distributed systems were discussed in Section 6.3.

As the future unfolds, the ways in which the system can actually be improved will inevitably change. New hardware models are launched. These can provide opportunities but of course they might also have the effect of making the existing equipment seem obsolete. Improved versions of the software may be issued and entirely new software modules for rather different applications can come on the market.

It is part of the responsibility of the computer manager to keep abreast of all developments in this field. General awareness can be enhanced by visiting exhibitions, attending conferences, reading magazines and other literature. Liaison with the supplier is clearly important and attendance at meetings organised by the user group related to the particular CAD system is always valuable. The user group is of course the means of meeting and sharing experiences with the persons in other user companies who are tackling common problems. Time must be allocated for these purposes.

Hardware upgrades might be proposed by the computer manager so as to increase capacity or otherwise to improve performance. For example some more workstations might be added to a successful system, an electrostatic-plotter might be substituted for the pen-plotter to speed up the output of drawings, or more memory might be added to the computer to speed up processing. In all cases, a carefully argued case must be built up and presented (refer Chapter 4). Clearly the CAD director must become convinced of the need, and indeed he will become the key individual in the decision-making process. Naturally it is always easier to get agreement to upgrade a successfully-managed system. Records which prove the success are always useful (refer to Section 5.12).

New versions of the software are issued from time to time. These may correct for software errors in earlier versions and perhaps provide some new or improved facilities. The new versions should be provided as part of the maintenance and support contract. The computer manager must see that these are installed, and arrange for access by the supplier's staff for this purpose.

Major enhancements to the software, or modules for new applications are another matter. These are rarely supplied as part of the support contract. Again if there is a need for them, the case must be built up and presented, as for new hardware.

CHECKLIST FOR IMPLEMENTATION – HARDWARE AND SOFTWARE

1 Plan and prepare the accommodation for the computer and plotter (where appropriate).

2 Plan and prepare the sites where the workstations are to be located.
   Options: Concentrate them in one area.
                 Disperse them within the design offices.
   (a) Daylight to avoid glare and reflections.
   (b) Artificial light to avoid glare and reflections.
   (c) Temperature control.
   (d) Seating.
   (e) Furniture and fittings.
   (f) Power and communications wiring.

3 Plan procedures, training and instructions for starting, stopping and operating main items of hardware.

4 Plan and institute procedures for:
   (a) Passwords for access.
   (b) Disc back-ups.
   (c) Archiving of project data.

5 Compile a disaster contingency plan.

6 Decide how the plotter will be managed, and organise consumable materials.

7 Carry out contingency planning for data exchanges that may be required.

8 Check preventive maintenance of hardware is being carried out regularly.

9 Set up logging schemes for:
   (a) hardware failures,
   (b) software failures,
   (c) support requests,
   (d) use of computer resources.

CHAPTER 7
Preliminary Project Design

This chapter reviews the use of CAD for preliminary, or conceptual, design of projects. The majority of CAD systems are not yet employed for this design phase, but I intend to show that there is much potential.

Site geometry has to be entered, this coming directly from survey equipment or from large-scale maps in digital format. Then models of competing schemes for the project can be built up within the computer. The merits of working in 2–D and 3–D format are discussed, and there is much to commend the use 3–D for this work phase. There is some potential for analysis of computer models to assist in scheme assessment. However, there are also difficulties and such analysis is not yet widespread practice.

Visualisation of 3–D models is however a powerful communications tool and it can help in assessment of competing schemes and in decision making. Various means of producing quality presentation material are discussed. The scope for working up preliminary design information for detail planning is also considered.

Preliminary design work is often undertaken by top-level designers. This does mean that arrangements may have to be made to tailor the CAD facilities so that operation by these people is possible.

7.1 PRELIMINARY DESIGN – TRANSFORMING IDEAS INTO A VIABLE SCHEME

Preliminary design is a term I shall use to denote the early design stages. In the terms of the RIBA Plan of Work, it would cover stages B, C and D (Feasibility, Outline Proposals, and Scheme Design). It covers conceptual design, when ideas are transformed into a viable scheme.

The design process normally begins with the production of the client's brief. This ought to be framed in terms of the needs of the situation, and of project objectives. The brief should not consist of firm or even preliminary ideas for solving the problem. These might tend to channel the designer's efforts into looking only at these suggested solutions. They might confine what in the early design stages ought be the designer's opportunity for a free-ranging exploration of the problem.

This preliminary phase is as critical as any part of the design process. We noted in Chapter 2 that there are innumerable solutions possible for any design problem. Since the design resources are always limited, successful preliminary design depends on quick-fire generation of good ideas. So creative ideas have to be generated, examined, assessed, evaluated, modified and communicated to others. Ideas are the commodity that any design firm needs in profusion if it is to thrive. Yet sometimes the manipulation of ideas and turning them into reality can be a long and frustrating process.

Can the computer help? Let us be clear that the majority of CAD systems are not employed in preliminary design. They are used for the later stages of detail layout and design, and for the production of design documentation. Some designers would even argue that computers have little or no part to play in preliminary design.

Nevertheless, I feel sure that CAD techniques can be used gainfully at most of the design stages, and the preliminary phase is no exception. So this chapter is devoted to an explanation of how CAD can supplement and improve on 'manual' preliminary design practice. It will also present a few examples of work that has already been carried out in this area.

7.2 A STOREHOUSE FOR IDEAS

The architect at this time will be thinking mostly about concepts of mass and space. What building form and layout will best meet the client's requirements? Apart from dealing with the client's wishes, there are also the needs of the future users of the building, the neighbours, the planning authorities and many others.

The structural engineer when appointed is concerned with the form of building skeleton which provides the strength and stability. The services engineer has to look at alternative means for maintaining a suitable internal environment, and at the provisions to be made for communications and drainage. The quantity surveyor or cost consultant is already having to balance the probable costs of proposals with the financial resources that will be available. A contractor or someone qualified to advise on construction aspects might well enhance the design team, but sadly such an appointment at this stage often seems just too difficult to organise.

The design tools have traditionally been no more that a pencil and drawing sheet. Yet these are far from satisfactory. Many drawings have to be prepared of competing schemes. It takes too long to prepare paper drawings after they outgrow the 'back of the envelope'. Often there is a need to start afresh with new drawings when significant changes are made to schemes. Designers soon tire of drawing and redrawing, and as a result there is sometimes insufficient incentive to keep the design evolution going. This is a serious matter because designers can be tempted to accept a scheme which they know is still some distance from their ideal.

The CAD system is an alternative technique for holding and manipulating design information. Computer memory is becoming a relatively cheap commodity. We should however remain realistic and not expect the computer to solve all the problems. We have to understand its limitations. The computer cannot produce any design ideas, but it can be a storehouse where ideas in the form of graphical, geometric and descriptive information are kept. We discussed the different forms of computer model in Chapter 3. Indeed several models can be constructed, each to represent a competing scheme for a project. Images can be produced on a screen which depict views of a model, and these can be plotted on paper when they are required as a record or for communication to others. The models can be modified as and when the design ideas evolve. It is the computer that does most of the hard work of regenerating the screen images and replotting the drawings after new modifications are incorporated.

This has been a brief summary of the computer's potential. We will now look in more detail at the various elements of this.

Creating and saving models

Computer-aided design tools are needed for:

(1) The input and storage of embryonic design solutions. This must be done in a manner which is quick and easy. Held in computer models, the information is then available for future reference and use.

(2) Manipulation and modification of these models.

If these requirements can be met, we have the means of keeping the design fluid while competing ideas are examined. Computer methods must enable more iterations of the design loop to be undertaken, not fewer. Any difficulty in handling the computer devices themselves must not inhibit too much the flow of ideas from the designers. Also the modelling facilities provided by the computer system must not inhibit the range of design solutions that can be examined. For example, some early computer aided design systems could cope only with certain fairly standardised building systems. This sort of limitation would certainly not be acceptable to most users today.

What financial benefits are available? At this preliminary design stage, we should not concentrate on looking for direct productivity gains from the computer-aided designers, measurable at any rate in terms of reduced manhours. There may well be benefits but they are more likely to appear as improvements in the quality of the end-product – in a better preliminary design. There could well be substantial indirect productivity gains too, but these are likely to be found downstream. By this I mean later during the detail design of the project, and later still when the construction phase begins.

7.3　SITE DATA AND CAD

In preliminary design, it can be difficult for the designer to know where to start. A blank screen, like a blank drawing sheet, can be daunting.

Every project has to be set in an unique landscape, so almost invariably some form of site survey must be put in hand. Clearly the nature of the project will determine what sort of survey is required. However, the method of computer representation to be adopted will also have a bearing on this.

Site surveys

In all cases, consideration has to be given to:

- the purpose of the survey;
- the information that is required, i.e. the type of features to be recorded;
- the required level of detail and accuracy of records;
- the manner in which the information is to be communicated.

In some instances, the site might be nearly level and the new project is intended to nearly fill the area. Then a few measurements only might suffice. At the other extreme, the site might be large, hilly or contain some special features. An important design objective might be to create a facility which will relate well to such an environment. There may be existing structures which have either to be demolished, avoided or included within the new scheme. In most cases there will be surrounding buildings which have to be considered.

Measured co-ordinates of survey points might be received in tabulated form, and these can be keyed into the CAD system. This is laborious and prone to gross errors. A survey plot supplied by the surveyor could be digitised, but this is still somewhat laborious and subject to lower-order, but significant, errors.

Surveys are carried out nowadays with electronic distance measuring (EDM) equipment or with electronic total stations. Readings and calculated results can be stored digitally by these so as to be directly available for use by computers. Interface programs are usually required to convert the data from the instrument readouts into the format required by a target CAD system. For example t^2 have linking software which can be used to accept survey data directly into their Sonata or Rucaps database. Such automatic transfer of data by an interface program is preferable over other methods, especially when a large volume of data is involved. The main benefit is the reduced probability of introducing errors, but time and effort can be saved too.

Information from large scale maps or plans of the area can be used to supplement the site survey. Most mapping agencies are turning to digital mapping methods. The Ordnance Survey in Great Britain [22, 23] has 225,000 sheets of 1:2500 plans, or 1:1250 plans in built-up areas, and there is a long, on-going programme for digitising these plans into computer format. By mid-1987 about 30,000 sheets had been encoded and were available on magnetic tape. Interface programs can read this data into the CAD database on a sheet-by-sheet basis.

Eventually a map user will be able to order most Ordnance Survey maps tailored to precise requirements. Thus buyers will be able to specify any plan extent or scale for their maps, the classes of information (or features) to be included and the colours in which they are to be shown[22].

Aerial photogrammetric methods of survey can be preferable for very large sites. Terrestial photogrammetrical methods, or photometrology[24, 25], might well prove to be useful for 3–D surveys of all the elements of existing large buildings or industrial plants.

7.4 VISUALISATION AND ASSESSMENT TOOLS FOR THE DESIGNER'S USE

Traditionally, the designer makes drawings to help to visualise the emerging design solutions. These drawings might well start as 'back-of-the-envelope' sketches. Before long they must graduate to carefully-produced drawings which incorporate much more detail.

Limitations of two-dimensional working

A 2–D computer drafting system could of course be used for preliminary design. We discussed in Chapter 3 how this method mimics the manual approach. It enhances drawing productivity and other benefits can be achieved. However, in the preliminary design phase, 2–D drafting does not meet some of the designer's foremost needs.

It is not too easy in practice even for experienced designers to look at two-dimensional plan drawings and then to visualise how the project will appear in its three-dimensional reality. Designers are usually far too confident of their ability in this respect. They rarely draw nearly enough sections, isometrics or perspectives. The reason is fairly obvious. The manual production of such drawings is a highly skilled task. It requires much time and effort to produce by hand even one accurate perspective view of a building scheme. Yet in the absence of such visualisations, many problems tend to be overlooked only to reappear during detail design. The real cost of such oversights is magnified as the time to their discovery mounts. It is far worse of course when the problems only surface on site, or after the completed facility is occupied.

When a 2–D computer drafting system is used in preliminary design, the design situation is improved little. The computer holds insufficient geometrical information to enable any program to automatically produce 3–D perspectives. Ideally therefore, we must build 3–D computer models during preliminary design.

Merits of three-dimensional modelling

In many 3–D computer modelling systems today, the means provided for viewing schemes are often fairly well developed. Visualisation of models means that the designer's subjective judgement can be brought into play. This is especially useful when the design objective is dominated by aesthetics. The 3–D model can become a most useful design tool to investigate alternative schemes for the project. Indeed the designer's ability to look at parts of his schemes from various viewpoints, to examine for clashes or all manner of potential problems, is indeed a powerful facility. This ought not be underestimated.

At this preliminary phase, the computer models may lack much in the way of detail. Indeed they become too cumbersome if too much information is fed into them at this early design phase. This defeats the object of the work. The acts of constructing and modifying the model would be slow. The generation of 3–D would require too much in the way of computer resources. All the benefits of using 3–D geometry would be nullified.

Fig. 7.1 is a view typical of what might be plotted at an early stage in a design project. This was produced in order to help the designer to visually assess the approach and main entrance of an evolving scheme. The view is a simple line perspective, rather than an elaborate colour-shaded view, because speed of model building and of viewing it is of the essence here. Line perspectives are generally preferred by designers when visualisations are required for their own use. The effects on the chosen scheme of design modifications – good or bad – can be seen more easily.

As the design work proceeds, model content and complexity can be gradually stepped up by a process of adding more components, such as staircases (Figs. 7.2 to 7.5). More detailed components can be substituted for simple components. At this time, the designer must search for the function subdivisions of the whole project into manageable sub-units, such as typical floors, service cores, toilet blocks, staircases and recurring components.

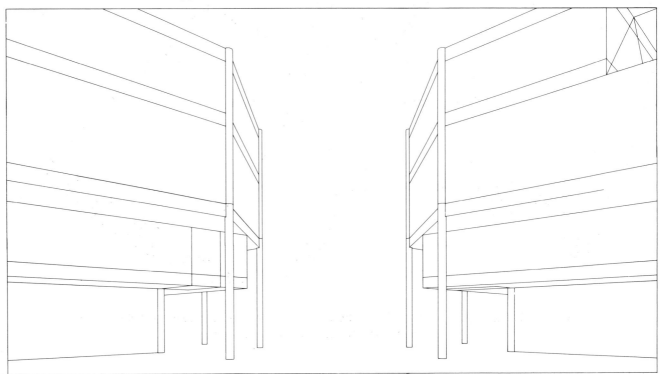

Fig 7.1 Microbiological Research Laboratories. Preliminary view of approach to the centre. (*Courtesy*: Baldwin Brattle Connelly Partnership, Stevenage, England.)

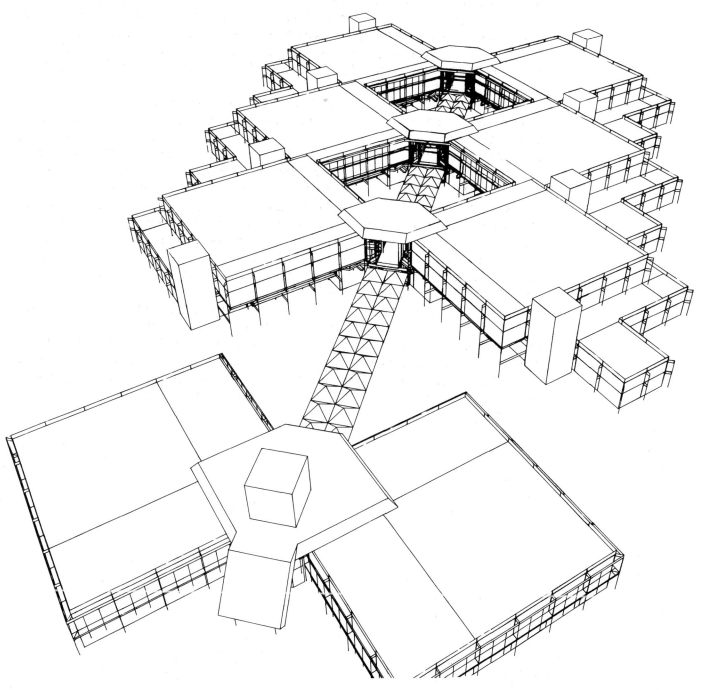

Fig 7.2 Microbiological Research Laboratories. Preliminary design. (*Courtesy*: Baldwin Brattle Connelly Partnership, Stevenage, England.)

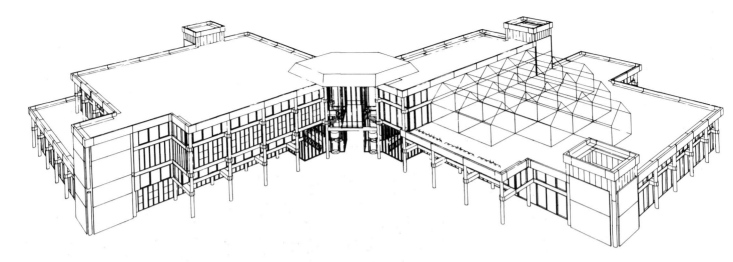

Fig 7.3 Microbiological Research Laboratories. Preliminary design. (*Courtesy*: Baldwin Brattle
Connelly Partnership, Stevenage, England.)

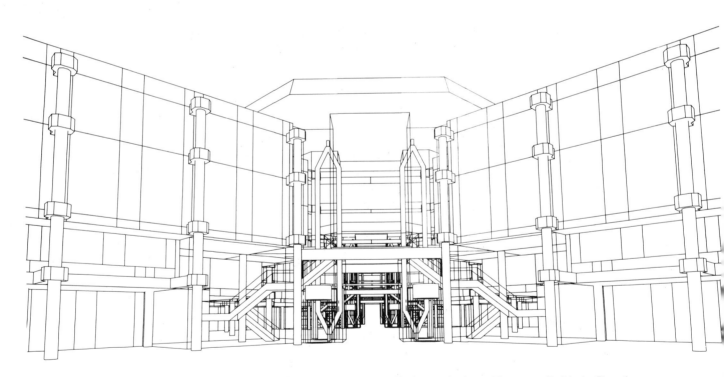

Fig 7.4 Microbiological Research Laboratories. Preliminary design. (*Courtesy*: Baldwin Brattle
Connelly Partnership, Stevenage, England.)

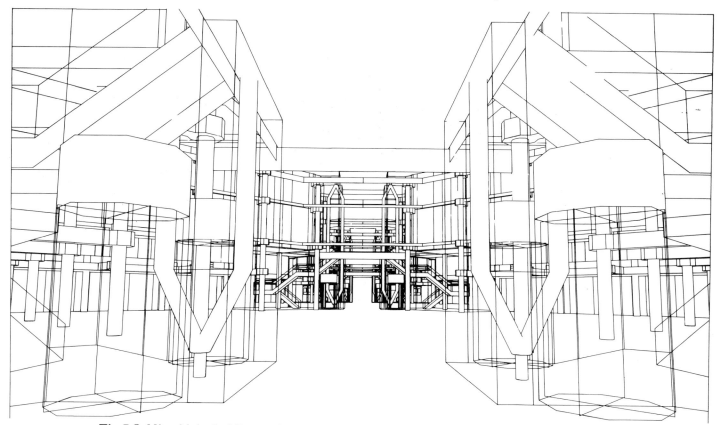

Fig 7.5 Microbiological Research Laboratories. Preliminary design. (*Courtesy*: Baldwin Brattle Connelly Partnership, Stevenage, England.)

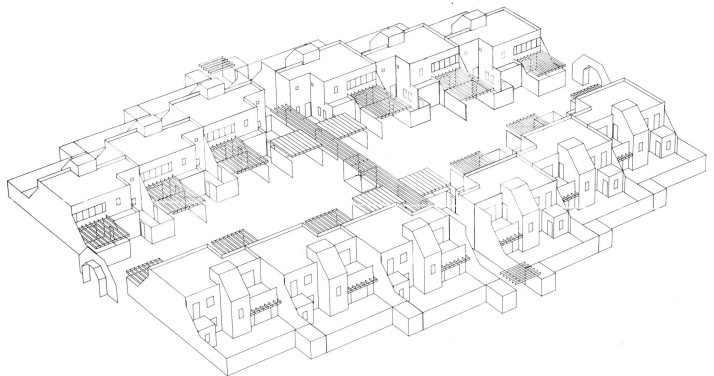

Fig 7.6 Housing. Preliminary design. (*Courtesy*: Baldwin Brattle Connelly Partnership, Stevenage, England.)

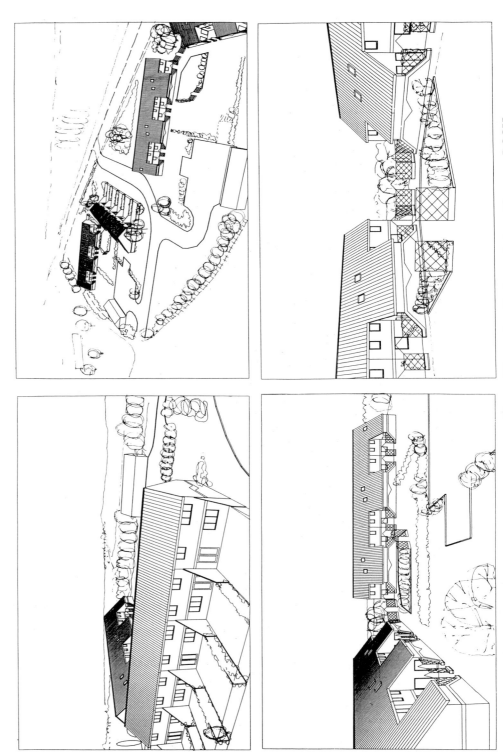

Fig 7.7 Servicemen's housing in Wimbish, Essex. Views of preliminary design. (*Courtesy:* BCCH Partnership, Welwyn Garden City, England for P.S.A.)

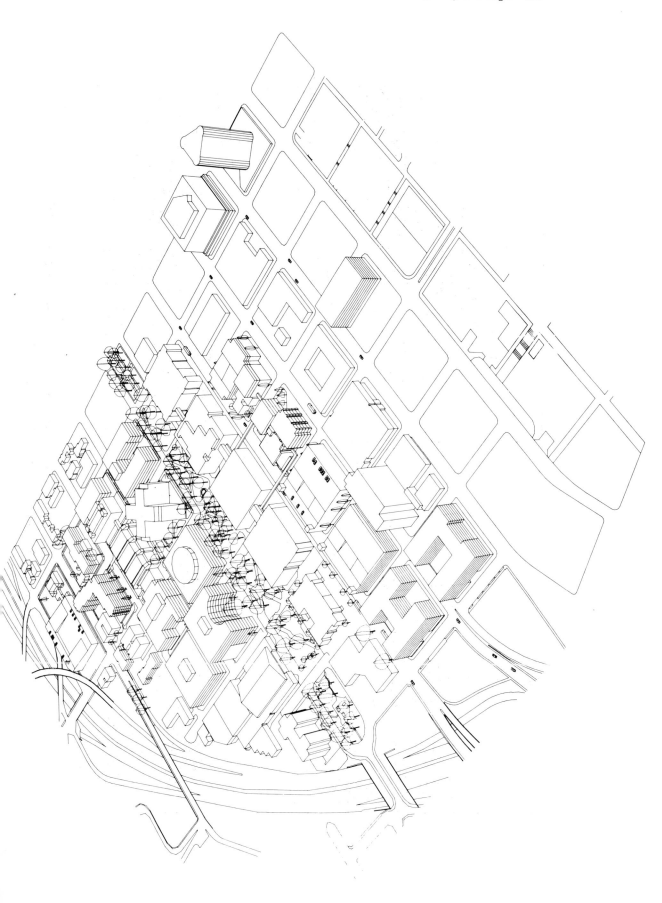

Fig 7.8 City plan. (*Courtesy:* Yost Grube Hall, Portland, Oregon, USA.)

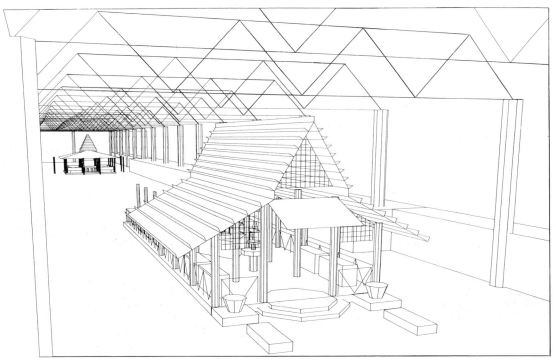

Fig 7.9(a) Proposal for restaurants, Bangkok Airport. (*Courtesy*: White arkitekter AB. Gottenburg, Sweden.)

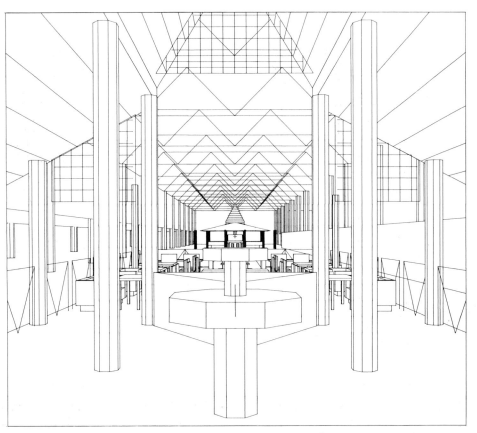

Fig 7.9(b) Proposal for restaurants, Bangkok Airport. (*Courtesy*: White arkitekter AB. Gottenburg, Sweden.

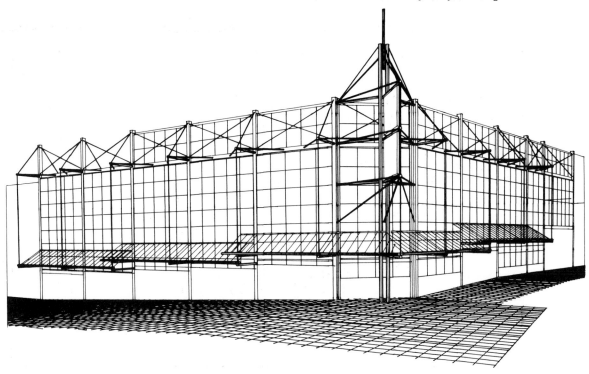

Fig 7.10 Shops and Cinemas, Sheffield for Wilson (Connolly) Properties Ltd. (*Courtesy*: Hadfield Cawkwell Davidson, Sheffield, England.)

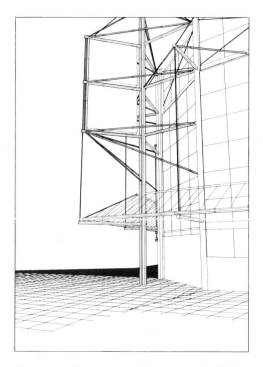

Fig 7.11 Shops and Cinemas, Sheffield for Wilson (Connolly) Properties Ltd. (*Courtesy*: Hadfield Cawkwell Davidson, Sheffield, England.)

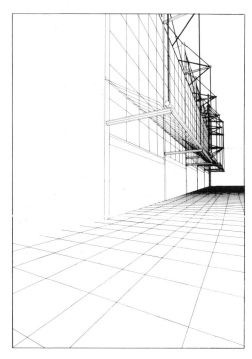

Fig 7.12 Shops and Cinemas, Sheffield for Wilson (Connolly) Properties Ltd. (*Courtesy*: Hadfield Cawkwell Davidson, Sheffield, England.)

Fig. 7.6 is a line perspective of a model where there was ample opportunity for repetition of a simple basic design. It may be seen that at this preliminary design stage, walls have been represented in an effective manner using simple surfaces only. Once the model has been constructed, any number of views from different viewpoints may be generated automatically (Fig. 7.7). These can provide the designer himself with a reasonable impression of how his finished project will appear.

Fig. 7.8 demonstrate how models of office blocks can be placed within the model of a site. The blocks can be rearranged until a satisfactory layout has been achieved. Figs. 7.9a to 7.12 are other examples of model-building where the main purpose was to help the designers to visualise their evolving design solutions. All illustrations have been prepared using 3–D models and so are samples of what can be produced. They were not produced as 2–D drawings.

Analysis of preliminary design models

Such perspectives are quite adequate for the designer's own visual assessment of schemes. Facilities for analytic assessment are another matter, and potentially these are important too.

A 3–D model contains the geometrical information and perhaps other attributes of the design elements as well. Obviously this is all in machine-readable format. Theoretically at least, it is available as a data source for any analytical computation or process that might be required. Thus it is easy to imagine how an analysis of the structure, or how some heat loss or energy calculations could be performed as part of the study of the design proposals. After any design change has been made, a re-analysis of the model could be initiated. Comparison of the results before and after the revision would help in the assessment of the merits of the design change. A 2–D model of the proposals is unlikely to be of much use for such analytical work. There is certainly more potential with the 3–D model.

This looks interesting and useful. There are plenty of examples of such anaytical applications. However the use of 3–D models for analytical purposes as yet is not widespread because it is rarely quite so easy in practice.

First the data must be in the precise format expected by the particular analysis program. This will hardly ever be the case, so usually a data conversion using an *interface* program is needed to fill the gap. Such interface programs are not always readily available. If anyone is to set about developing an interface, then the data structures of the CAD system and the analysis program must be understood. Development of interfaces is often inhibited by the fact that CAD system suppliers are loathe to release to anyone the precise structure in which the design data is held within their system.

A further problem is that the CAD model may hold much information useful for an analysis program, but it rarely holds all the necessary data. For example, for a structural analysis we need more than the geometry of the structural outlines. The structural engineer prefers to analyse some idealised structure, perhaps an idealised 2–D or 3–D frame, rather than the real structure denoted by its precise outlines. The analytical program needs additional information relating to structural assumptions, loading, material properties, and other items that will not be held in the CAD model. A similar situation would apply with energy or other analyses for a building.

The conclusion we must come to in respect of visualisation and assessment of models is this. There is indeed potential, much of it at present unrealised, for analysing and assessing 3–D design models held within the computer. The production of useful and realistic-looking views of models is already a practical proposition.

Plate 1. Competition entry for Financial Services Centre, Georges Dock, Dublin. (*Courtesy*: Real Image, Welwyn Garden City, England and Arthur Gibney & Partners, Dublin Eire.)

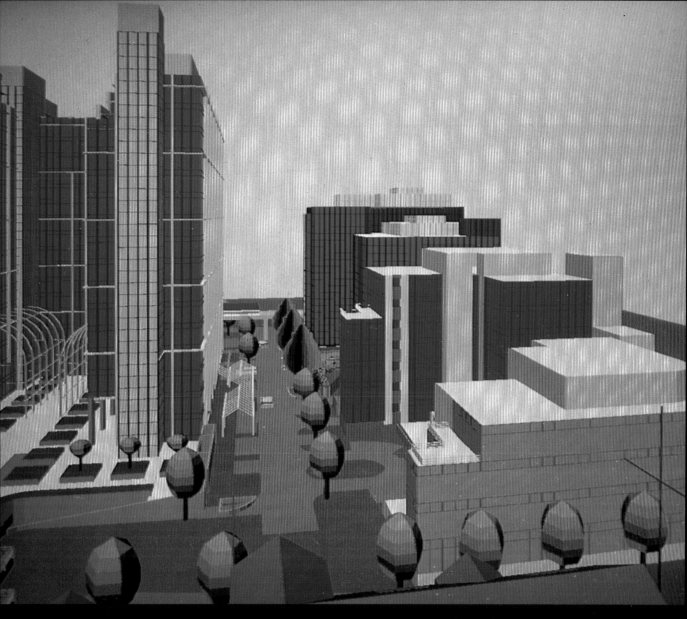

Plate 4. *Above.* Harbour Exchange. (*Courtesy*: Real Image, Stevenage, England and Frederick Gibberd Coombes & Partners.)

Plate 2. *Opposite above.* The Pool. (*Courtesy*: Guliano Zampi. D Y Davies Computer Services, Richmond, England.)

Plate 3. *Opposite below.* The Only Move to Make (*Courtesy*: Jo Templin. Oldham Boas Ednie Brown. Perth, Australia.)

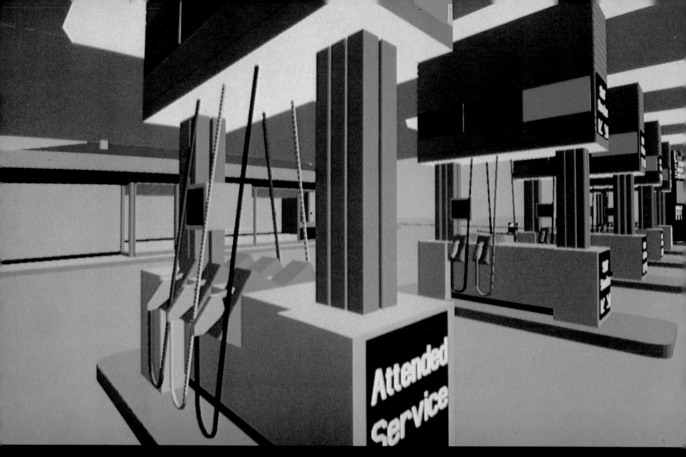

Plate 5. Shell filling station (*Courtesy*: Shell UK Oil Retail Division, London.)

Plate 6. Refurbished Wool Exchange, Melbourne (*Courtesy*: The Building Modelling Company Pty Ltd, Architects, Peddle Thorp & Learmonth, Melbourne, Australia.)

Plate 7. The Rucaps Design Office. (*Courtesy*: Robert F. Paolini. Coles Myer Ltd, Melbourne, Australia.)

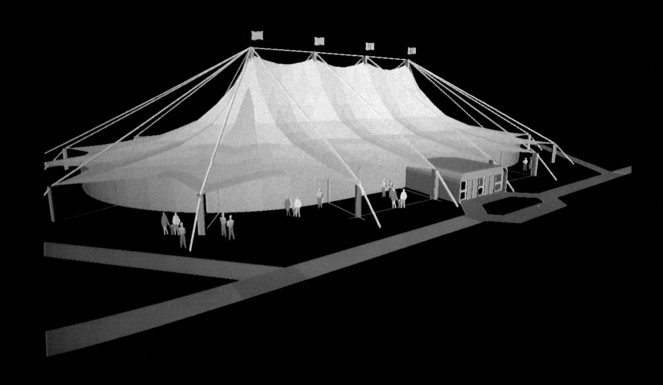

moss
moss

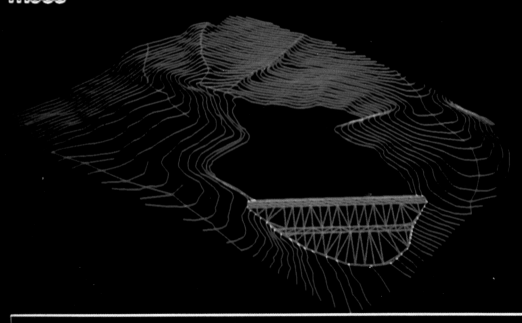

RIVER, MARINE AND WATER ENGINEERING
- LAND AND HYDROGRAPHIC SURVEYS
- PIPELINE ROUTEING, PLANS AND SECTIONS

Plate 11. *Above.* Melbourne office development. (*Courtesy*: The Building Modelling Company Pty Ltd. Architects. Peddle Thorp & Learmonth, Melbourne, Australia.)

Plate 9. *Opposite top.* Tented Structure. (*Courtesy*: Norman Mitchell. t² Solutions Ltd. Berkhamsted, England.)

Plate 10. *Opposite below.* Reservoir modelling. (*Courtesy*: Moss Systems Ltd. Horsham, England.)

Plate 12. *Overleaf.* View of Central Courtyard. (*Courtesy*: Scott Brownrigg & Turner, Guildford, England.)

7.5 COMMUNICATING PRELIMINARY DESIGN INFORMATION

Preliminary design incorporates lots of ideas, and many different individuals become involved with their generation, assessment, and with the ensuing decision-making. Clearly the communication channels among all the parties must be efficient.

Design communications are never easy. A stranger to the construction design process might perhaps come to the conclusion that a client's interests would be best served if all the experts were in one tight-knit organisation. This could be a multi-discipline design firm or else a construction company having its own design office. However, the construction industry does not always work in such a logical manner.

Communications within the design team

At present, the normal medium of communication is the drawing print which can be delivered by hand or mailed. The architect tries to foresee what information the structural engineer will need, and incorporates this in his drawings. Later the engineer will send information back to the architect. Neither of the parties is completely sure of the needs of the other. So the drawings communicated are all too frequently the products, like floor layout plans, that have already been produced for the originator's own in-house requirements. As a communications medium they can be unsuitable, because they may contain either insufficient or superfluous information for the intended recipient.

In Chapter 8 when we come to look at project management, we shall see that information exchanges ought to be carefully planned before they take place. Then the various parties are more likely to receive the required information at the time when they expect and need it.

At any time, a drawing sheet showing the current proposals can be requested from the computer and plotted. This will be done when there is a need for a more permanent record, or a need for communication of a scheme to others. Each of these drawings is newly generated and clean, and is a by-product from the modelling process. Indeed it can be practical to generate sets of unique drawings from the design information held in the computer. Each sheet can contain the specific classes of information required by each recipient.

We should not forget schedules. We shall see in later chapters how non-graphical design information can be accumulated in some CAD systems. Then schedules of accommodation, of room areas and even cost-plans can be generated at the preliminary design stage.

Communicating with the client

Let us now look at the special problem of communication of scheme designs to clients. Most clients lack the ability to visualise from floor plans alone what it is that they are being asked to pay for. Effective communication often breaks down. All too frequently the real buildings which eventually appear come as something of a shock to clients. The reality is different from what they had visualised in their own minds. The shock is all the worse because they are powerless at a late stage to do anything about it – except to make a resolution that another design team will get the next commission.

With a computer modelling system in use, 3–D views like that shown in Fig. 7.13 can give the client an excellent impression of the internal layout of the design proposals. Views of competing schemes could be circulated for comment too. These can supplement or even replace the traditional floor plans that are currently exchanged. Moreover, such 3–D views ought to be fairly freely circulated to all concerned, so that

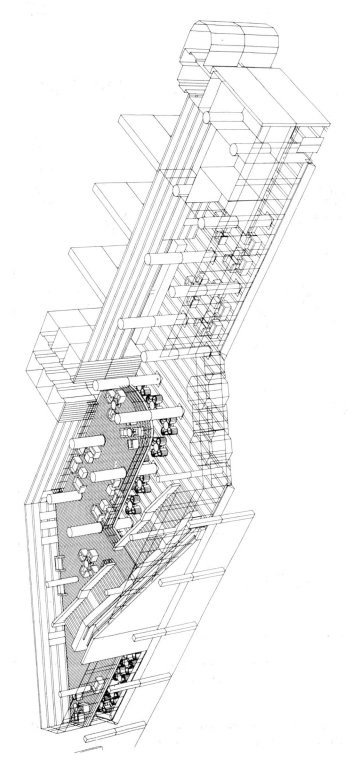

Fig 7.13 Proposal for a hotel building. (*Courtesy:* White arkitekter AB. Gottenburg, Sweden.)

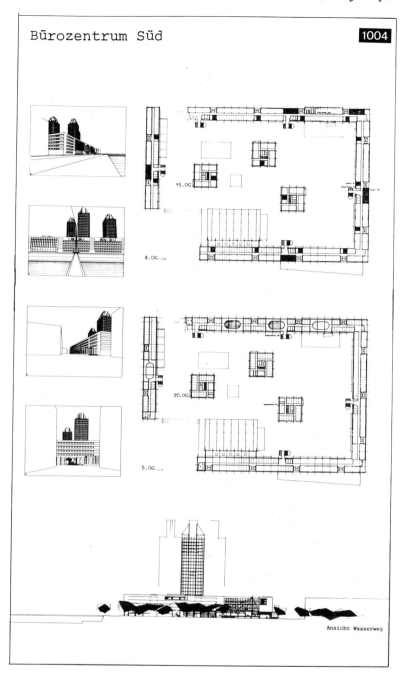

Fig 7.14 Burozentrum Sud, Frankfurt. (*Courtesy*: Heinle Wischer & Partners, Stuttgart, West Germany.)

proposals can be more open to inspection and critical evaluation. It is also possible to compose a drawing with a mixture of 2–D and 3–D views (Fig. 7.14).

7.6 DECISION-MAKING AND CONTROL OF DESIGN PROCESS

Initially the design proposals must to be kept fluid, while competing ideas are studied. Having properly examined the schemes, the team must solve all major problems associated with them as soon as possible. Then of course a time must come when the process is 'frozen' with a generally-agreed solution. If the problems are not properly solved early on, then quite simply the team could proceed on into the detail design with the wrong design solution.

In whatever way we undertake preliminary design – traditional means or CAD – the function does need to be project-managed. Someone has to be in control of the exchange of design information. With the design proposals evolving so rapidly, someone has to know which of the proposals is the preferred solution at any time. If this is not properly looked after, people are going to waste time working on solutions which are already out-of-date. Project Management with CAD is the subject of the next chapter.

7.7 PRESENTATION OF SCHEMES

When the design professionals are agreed on a solution, they usually have to present or 'sell' their ideas. This requires communication and persuasion. High-quality presentation of ideas can be particularly important with prestige projects or those which will have much impact on the public at large.

Presentation of design proposals might have to be to the client, the future users, a planning committee, or to the general public. There might even be a formal public enquiry. These parties all tend to have one thing in common. They usually lack design ability and the designer's insight. They are lay-people. They have little ability to grasp abstract concepts of building design. However they are intelligent people and they know what they want when they perceive it.

Selling is difficult enough when some real or 'concrete' product is on hand to be demonstrated, tested and looked at from every angle. Design proposals by contrast are no more than ideas. This is why traditionally we often have to present one or two expensive hand-produced drawings like perspective views worked up by artists. Frequently resort has to be to costly physical scaled-models.

At this preliminary stage the design team might still be working in direct competition with other teams. This adds an extra 'edge' to the selling process. We must be careful that mere superficial quality of presentation achieved by the opposition does not win over the design quality of the 'home' team's proposals. In short, we need to win with design quality plus presentation quality.

A range of presentation methods will be reviewed.

Line perspectives

After a 3–D CAD model has been created, we know that it is a fairly straightforward matter to generate line perspectives from many viewpoints. Used imaginatively line perspectives can give a good impression of the final appearance of a building (Fig. 7.15).

Fig. 7.16 is a wireframe view of the courtyard which is to form the central feature of a large building. Hidden lines have not been supressed so that the structure and layout is clearly apparent. Fig. 7.17 features the connections from the exterior of the building. Fig. 7.18 is a wireframe view of the whole scheme, again with no supression of hidden lines so that the courtyard, entrances and connections can be clearly illustrated. Fig. 7.19 and Plate 12 show an entrance and courtyard views. The model can even represent the entasis on the columns.

Once a model exists, a range of line perspectives can be produced fairly easily. The designer can quickly sift through them looking for ideas for presentation. Perhaps he can pick out a few of the more promising examples for further treatment.

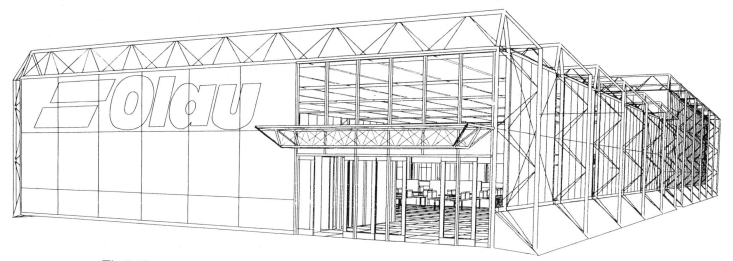

Fig 7.15 New Passenger Ferry Terminal, Sheerness. (*Courtesy*: Blue Chip Computer Services. Dartford, England.)

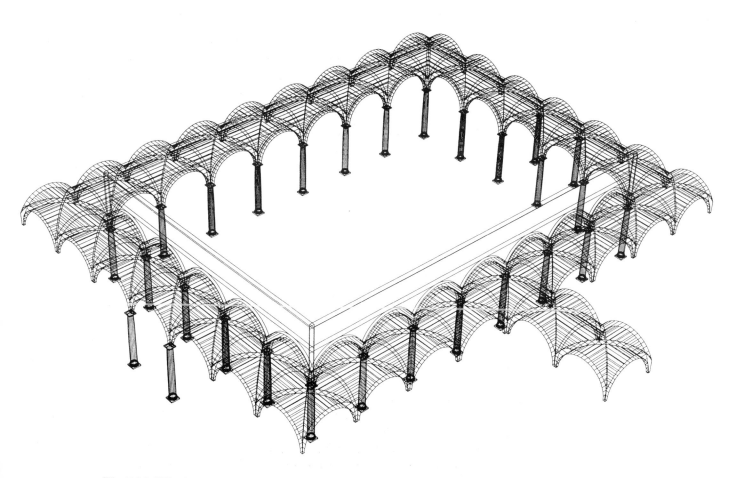

Fig 7.16 Wireframe view of central courtyard. (*Courtesy*: Scott Brownrigg & Turner, Guildford, England.)

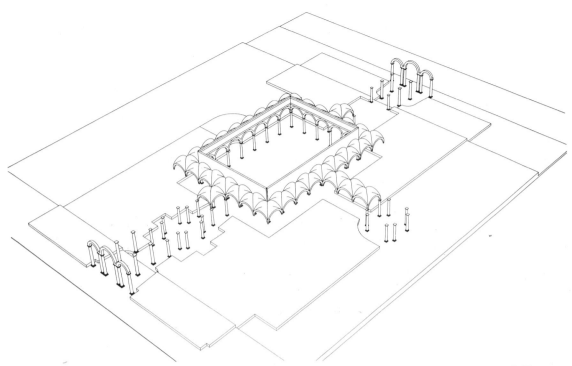

Fig 7.17 View of courtyard and connections to exterior. (*Courtesy*: Scott Brownrigg & Turner, Guildford, England.)

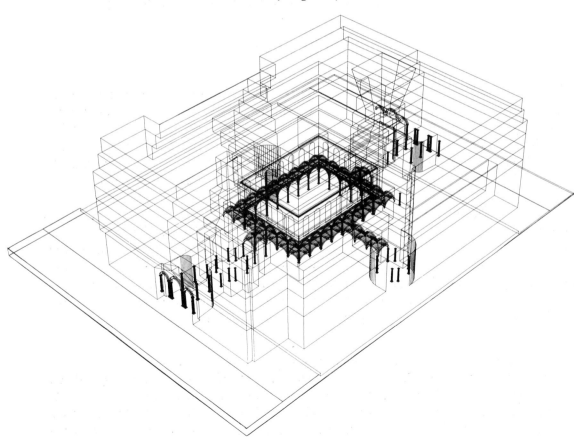

Fig 7.18 Courtyard, entrances and connections highlighted within building. (*Courtesy*: Scott Brownrigg & Turner, Guildford, England.)

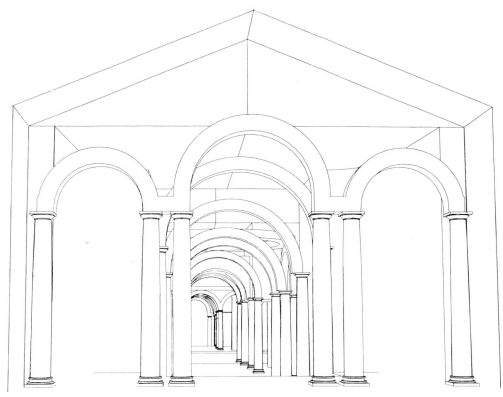

Fig 7.19 View of entrance. (*Courtesy*: Scott Brownrigg & Turner, Guildford, England.)

Shaded-views from surface models

Shaded-views can be made from a surface model. Three views generated from the model of the Financial Services Centre in Dublin are included in Figs. 7.20 to 7.22 and Plate 1. City-scapes in Melbourne, Hong Kong and Glasgow are shown in Figs. 7.23 to 7.25.

With suitable hardware and software, the results can be produced in full colour. Even when appropriate hardware or software is not available in-house, the design model might be passed over to a computer bureau and their assistance sought for the presentation output.

The results can be in the form of photographic prints, slides or as output from electrostatic, ink-jet or other types of plotters. Photographs taken in a blacked-out room of the graphics screen can of course only be as good as the pixel display of the screen. Sometimes the cathode ray tube distortion can be a problem. Several of the Plates provide examples of what has been achieved by this method. With suitable photographic equipment the display screen may be by-passed altogether, thus avoiding the distortion and limited resolution, by directly writing the image to film.

Slide presentations and audio-visual treatment

A number of colour slides can be used to accompany a verbal presentation by the designer. Alternatively, static pictures can be compiled into a pre-recorded synchronisation of film images and soundtrack, to produce an audio-visual feature. Arrays of slide projectors using dissolve techniques and synchronised with sound on cassette tape players can be used. These provide a more elaborate and highly effective presentation. Such techniques have been described by Watkins[26].

Fig 7.20 Financial Services Centre, Georges Dock, Dublin. (*Courtesy*: Real Image, Welwyn Garden City, England and Arthur Gibney & Partners, Dublin, Eire.)

Fig 7.21 Competition entry for financial Services Centre, Georges Dock, Dublin. (*Courtesy*: Real Image, Welwyn Garden City, England and Arthur Gibney & Partners, Dublin, Eire.)

Fig 7.22 Competition entry for financial Services Centre, Georges Dock, Dublin. (*Courtesy*: Real Image, Welwyn Garden City, England and Arthur Gibney & Partners, Dublin, Eire.)

Fig 7.23 Melbourne City model. (*Courtesy:* The Building Modelling Company Pty Ltd Architects. Peddle Thorp & Learmonth, Melbourne, Australia.)

Fig 7.24 Hotel complex in Hong Kong. (*Courtesy*: Jardine Engineering, Hong Kong.)

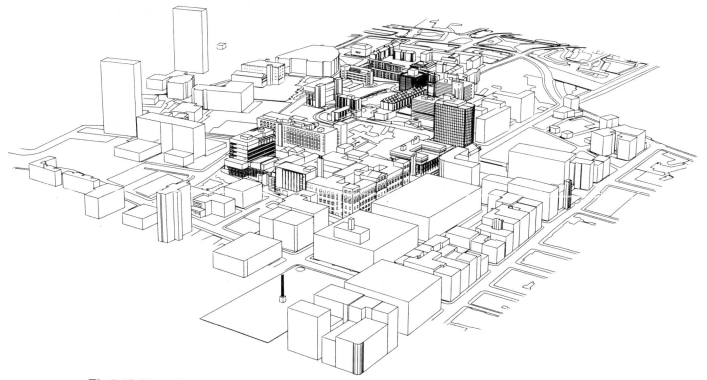

Fig 7.25 View of computer model of Central Glasgow. (*Courtesy*: Dr A.H. Bridges, Abacus Unit, University of Strathclyde, Glasgow.)

Animations

Animated video or film can be produced. For this, a large number of colour-shaded views have to be generated by the computer. Each view must be from a slightly different viewpoint, set at successive points along some suitable path. Panning of the animated view, and zooms in or out can all be devised. Stringing these together to produce the animation can be done if necessary by an external bureau or agency.

Where the design model contains the main elements of the external envelope of the building, like walls, roof, windows, doors, etc., then interesting walk-pasts or fly-pasts of a project can be created[27]. When the visually-important internal elements of the design are contained in the model, like partitions, columns, floors, ceilings, windows, doors and furniture, then walk-through animations can be produced. Abacus software has been used to provide walk- or fly-through views of Glasgow's streets[28].

Such animation can be a highly effective presentation medium as the advertising and entertainment industries already indicate to us. An evening's TV is peppered with computer animated 'spots'. Nevertheless, the resources required for their production should not be underestimated. First, a model with adequate detail must be created. Then each shaded perspective view requires a significant amount of computer processing time – perhaps of the order of ten to thirty minutes. Hundreds or even thousands of such views are required for the whole animation. Professionally-produced animated TV title sequences or commercials can cost over £100,000 for a fifteen to forty second product[29]. Such techniques are also at the heart of real-time graphics simulations used for flight training for pilots[30].

Nevertheless it is possible to produce a useful animation sequence as a by-product of design models at a much lower cost. One means is to extend the animation considerably by including static images.

Photomontage

Photomontage techniques are used when still perspective views of the design model are superimposed on photographs of the background (and foreground). Clearly the camera position must be noted or measured so that the correct view of the model can be generated. The computer output can be plotted on acetate sheets. These are overlayed on to the site photographs, and the completed montage again photographed (refer Fig. 7.26). The end-result can be a much improved visual impression of the intended project[31].

These techniques have been used by Abacus to investigate the visual impact of major industrial schemes on the environment. With more sophisticated equipment in the form of video editors now available, much can be achieved by techniques of this nature.

Artistic treatment of computer line perspectives

In many respects the wholly computer-based solutions already described still lack flexibility. There is sometimes value in combining the computer-produced line perspectives with manual treatment by an artist.

A few of the best or most interesting perspectives can be plotted and handed over to the artist. First of all the act of selecting a few from many generated by the computer helps considerably. Second, the arduous and time-consuming manual drafting of perspectives is avoided. Third, the result produced by the computer is perhaps more likely to be technically correct than manually-drafted products.

At this stage however the artist can take over and add features of the foreground and background. The view can be embellished with more far detail than exists in the computer model. The artist can add items like human figures, motor-cars, street

Fig 7.26(a) & (b) Photomontage showing impact of proposed apartment development. (*Courtesy*:
Real Image, Welwyn Garden City, England for Malshof Developments Ltd.)

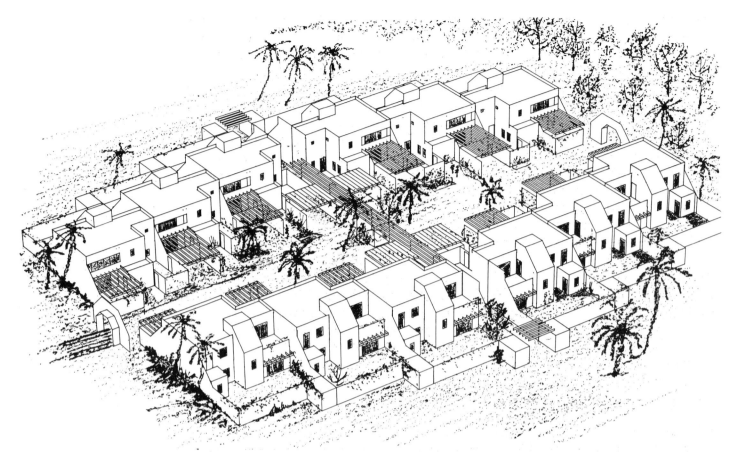

Fig 7.27 Artistic treatment applied to computer perspective (refer to Fig 7.6). (*Courtesy*: Baldwin Brattle Connelly Partnership, Stevenage, England.)

furniture, and vegetation. Besides giving a better impression of how the project will really appear within its setting, these help to give a feel for the true scale of the proposals. An artist too can emphasise or even highlight certain elements, while playing down others. There may well be merit in this, provided of course that too much artist's licence does not undo the accurate start provided by the computer.

In practice, the artist can soften the hard line of the computer perspective. The computer perspective shown in Fig. 7.6 was given artistic treatment. The result is shown in Fig. 7.27.

7.8 PROGRESSING IN PRELIMINARY DESIGN AND BEYOND

In the manual design process, some drawings may be produced to support the feasibility study or outline proposals. These normally have to be discarded when the scheme design stage starts, because then a new and more elaborate set of drawings is begun. In turn, these scheme drawings are discarded too when the detail design starts, because by this time a whole new set of drawings is required. Often these are done to a larger scale.

Computer model building is likely to be a less wasteful process. There is scope for a continual build-up of design information within the model. All design information is inserted at one scale, namely full-size. Initially for example, we might start with a 3–D model of the ground surface. The walls and roofs of a proposed design solution might be represented by simple surfaces in a model, while doors and windows can be represented by very simple CAD components. By this means, the preliminary models are kept in a fairly elementary form so that manipulation can be rapid. This way the models are more likely to keep up with the designer's rapidly evolving ideas. Drawings are extracted to serve initial purposes.

As time passes, the models evolve. The geometry is gradually adjusted and hardened up. CAD components which contain more detail can be substituted for the initial simplified or schematic versions. Cavity brick walls with the correct thickness are substituted for plane brick-coloured surfaces. Windows and doors representing and looking more like real products are chosen and placed. Problems of fit are encountered and solved. As the work proceeds, views of the evolving design solution are continually available on the designer's screen for critical examination.

Later still, having hardened on one scheme, the project is ready to enter the detailing stage. Assume at this time we have a 3–D locational model that was built up during preliminary design. This will contain 3–D representations of the major components and elements of the project – but not minor components or much detail. The accurate 3–D project geometry will be fixed. In the detailing stage it may not be a practical proposition to continue adding more and more detail in 3–D format. However, 2–D detail could be readily added to the 3–D framework as design work progresses.

So we can see that the earlier preliminary design effort can be built on. It does not have to be abandoned and a fresh start made. We will take up this subject of detail design in Chapter 10.

7.9 SENIOR DESIGNERS AS OPERATORS

I discussed the selection and training of CAD operators in Chapter 5. There are however some special considerations which relate particularly to preliminary design.

Normally it is the fairly senior people who have a role in preliminary design. It is not too clear whether it is merely a long experience of design work that seems to be necessary, or whether the most successful designers gravitate to senior positions fairly rapidly. Either way, we need to take into account that it is senior people, perhaps even at director or partner level, who undertake much of the preliminary design work.

If these people do all their design work through intermediaries such as technicians or draftspersons, then the all-important flow of ideas can be impeded. Some of these people do indeed like to work directly on the drawing-board. This applies particularly to senior architects.

If they transfer to computer modelling technology, then it follows that these seniors ought to be operating the graphics stations. Ideally they should be able to work up schemes until they become sufficiently detailed and settled. After that, the schemes could be handed over to the project architects or engineers for further design.

There can be a problem in this scenario! In some cases at least, it seems to be below the dignity of senior people to be seen touching a computer keyboard. This might be partly an irrational fear of a complex machine. Coupled with this there is their realisation that they cannot spend as much time as their junior colleagues at learning all the complicated operating procedures and commands. They do not like to be seen to be at a disadvantage to others.

For such senior people, there is today at least a ray of hope. Complicated machines can be driven by ordinary mortals. After all, high performance motor-cars can be driven along motorways – even without the services of a chauffeur. This is because motor-car operating procedures are nowadays fairly straightforward, after a little practice. Our machines can be driven hard with confidence that the journey's end can be reached without breakdowns, and without anyone even having to go and tweak the engine. Indeed it is increasingly true that the benefits of much complex machinery are nowadays available, without the users having to understand every aspect of their internal workings.

Unfortunately computer modelling is still in its infancy. There is still some distance to go before worthwhile modelling facilities can be rated as being simple to operate. However the situation is improving all the time.

In the meantime, it does not necessarily follow that each and every operator needs to be completely proficient at every aspect of operation. It must be emphasised that a CAD system can be implemented and people organised so that specialised tasks and responsibilities are spread among them.

For example, let us look at the operation of the system plotter. There are a few complexities involved in making sure that an ink-drawing is drawn satisfactorily. However, someone can be delegated to look after this machine as a specialist part-time activity. Our senior designer who is undertaking preliminary design work with the system should at least be spared from having to worry about this detail. The system must be organised so that he is able to order up a drawing, with some confidence that it will soon appear on his desk.

Likewise, our senior designer perhaps does not need to know how to create complex CAD components. An armoury of useful components should exist within a system library. It might be enough for our senior to be able to call up existing components, and to place these anywhere he requires them. What is important is that he can do creative design work with the aid of the system.

In short, one or two seniors should be able to use the system. They may need special support. Essentially the sysem must be organised for them so that it appears like a relatively easy-to-use 'black box'.

CHAPTER 8
Project Management

Chapter 5 dealt with the implementation of a system and the setting up of CAD management for the design office as a whole. Chapter 6 discussed the management of the system itself. This chapter moves on to consider the management of design projects. I do not intend to cover all the techniques of project planning and control. Rather I will discuss how design management should differ when CAD replaces the manual methods. The techniques actually employed inevitably must be varied to suit the size and complexity of the project in hand, and to some extent to match the personalities involved.

The key relationship which will be under scrutiny here will be that between the CAD co-ordinator and the managers of individual projects.

First the need for project management will be raised. Project planning for preliminary design and for detail design is separated out. For detail design, the planning can be done with the aid of bar charts and resource histograms, or by using more sophisticated network planning techniques. The demand for work-sessions at the CAD stations must be estimated. This can be expressed in the form of histograms. For multi-project planning, these histograms can be consolidated and then adjustments made to try to smooth the demand.

CAD should be applied selectively and a section deals with certain characteristics of projects which would be pointers to successful application of CAD methods. The chapter ends by considering the various agreements that need to be established among members of a design team for a project. A checklist for project management is included at the end.

8.1 THE NEED FOR PROJECT MANAGEMENT

When a project reaches the construction phase, the contractor has to organise and control all the workforce, materials and plant that will be required. The number, variety and expense of these resources can be considerable. They will be provided by the contracting organisation itself, and by its many subcontractors and suppliers. In these circumstances it is usually well appreciated that the project must be very carefully managed if it is to be completed profitably, within time-limits and to the required quality.

The design phase is no different in principle. Certainly the management problem is a little simpler compared with construction management. This is true because the number of parties involved is smaller. Also few materials and plant are needed.

The design management may be easier, but this is certainly no reason for ignoring it. The construction process is critically dependent on receiving the design information. If this is late, insufficient, erroneous or subject to later revision, then of course the effect on the construction work can be very expensive or worse.

In Chapter 5, I complained about the dismal standard of design-office management that is so prevalent today. I argued that there are many important reasons why this state of affairs is unlikely to be tolerated for much longer. Construction projects are tending to be larger and more complicated. The design team is getting larger too. There are more statutory requirements to be met. This means that an increasing number of exchanges of information has to be made during the design process, and approvals to be gained. There is more pressure to complete the design work as early as possible. The cost of design staff and office accommodation is rising all the time. To make matters worse, now there is far more fee competition and other forms of stress.

All this applies whether the design is done using traditional methods or with the aid of computers. Manual design techniques are thought to be established and understood – though this is itself in doubt. However, computer use, which implies a change to what initially are unfamiliar methods, certainly requires direction. The function of management is to provide the direction for a business and its parts. It follows that the introduction of a rapidly-developing but unfamiliar technology like CAD must increase the pressure – and need – for management.

It seems that the recognition of the need for project management, and a strong commitment to it, is of first importance. Compared to this, the techniques adopted are just a little less critical.

8.2 PROJECT PLANNING

In any endeavour, it is important to define what has to be done. Then we have to decide who can undertake the tasks, when each of these can begin and end, and what other resources including information are needed. The planning must take into account the work to be done by other organisations and the interactions necessary with these.

First we must step back a little. As soon as the project even becomes likely, some thought ought to be given as to whether it is going to be a CAD job.

We should be clear that some projects, or parts of projects, are not very suitable for the CAD approach. These are better tackled by normal manual methods. Secondly, in most organisations there is not enough installed CAD capacity to tackle every project. So the suitability of each project for the CAD method has to be assessed and a choice made. This is a subject which I will deal with in Section 8.6.

The plan itself must evolve as time passes, because of course the nature of the design process itself is dynamic. The initial plan will be tentative. It will however develop in an iterative way as the requirements gradually become clearer. The RIBA Plan of Work is one description of the administrative framework under which design projects can be undertaken. Perhaps however, this places too little emphasis on the iterative nature of the process.

Planning of preliminary design

In Chapter 7, I covered the use of CAD systems for the early stages of design and showed examples of what could be done.

Let us assume that this early design phase is to be tackled with CAD. The design solution at this time may seem to be too tentative for much in the way of formal planning techniques to be instituted. Nevertheless what must be done is to:

(1)   Define the objectives or goals for the design work.
         i.e. work content.

(2)   Construct a programme for the preliminary design work.
       i.e. timing of project design.
(3)   Fix dates for when the modelling work is to be started and finished.
       i.e. timing of CAD work.
(4)   Decide which operators will be involved. Decide also the probable number of workstation sessions that will be necessary.
       i.e. CAD resources.

As discussed in Chapter 7, sometimes when preliminary design models are being investigated it is better if highly experienced designers can operate the stations. At the planning stage we must ask ourselves if the relevant people are already trained to an adequate standard. If a designer is not very competent at using the computer system, will an operator who has had more practice be near at hand to assist? Are suitable CAD components already created for the preliminary modelling – perhaps in 3–D – so that model-assembly can be started? Several models may have to be built for investigation and comparison before a choice between optional design schemes can be made.

In planning this work, close collaboration is clearly necessary between the CAD co-ordinator and the manager or designer responsible for the project.

Planning of detailed design of project

When this phase begins, much more is known about the chosen design solution. The design task that lies ahead is better understood. From now on, one or more design models must be built up to represent the increasing detail of the planning. Drawings and schedules have to be plotted or printed. Design information will have to be exchanged between all the corporate bodies involved in the work. The scale of the work, measured in terms of design resources required, is now greater and careful management is correspondingly more important.

Usually the best planning approach is first of all to make lists as follows:

(1)   List the main items of information expected *from* other parties – with target dates. The parties typically are the client, survey firms, other members of the design team. The items will mostly be drawings and schedules. The list of information items should note the subject matter or contents of each document, and the party involved.
(2)   List the discrete problems to be solved or key decisions to be made. Again note the target dates, the parties involved, and a description of the nature of the problems or decisions.
(3)   List the drawings and schedules that must be produced in-house with the CAD system for *issue* to others – with target dates. Here too, the subject matters or contents of each should be noted. This list is the first draft of the register of drawings and schedules for the project.

This already constitutes the beginnings of a plan for the detailed design work. All the main tasks and key dates are identified.

The CAD co-ordinator will have to determine from this the nature of the CAD resources required. These are in forms such as:

- CAD operators,
- workstation sessions,
- items of design information to be gleaned from other projects.

8.3 PLANNING CHARTS AND RESOURCE HISTOGRAMS

The problem now is to rearrange all this information into some logical order. Selecting the right tasks to be tackled at the appropriate times will obviously ensure that the design work is effective. Good communications are the key to project management and so it is essential to make the plan presentable. Then all parties involved will know exactly what is expected of them.

Graphical presentation is far more acceptable to busy managers than tabulated reports.

Examination of the list of information expected from other parties (point (1) above) will identify some of the restraints on the commencement of in-house tasks. Potential bottlenecks might show up.

An analysis of the problems and decisions (point (2)), and of the in-house documentation to be produced (point (3)), will assist in identification of the design resources required. The nature of the documents required will throw light on the nature of the design information that must be assembled into design models, and when this is necessary. This in turn indicates the CAD components and other graphical information that have to be created or transferred from elsewhere.

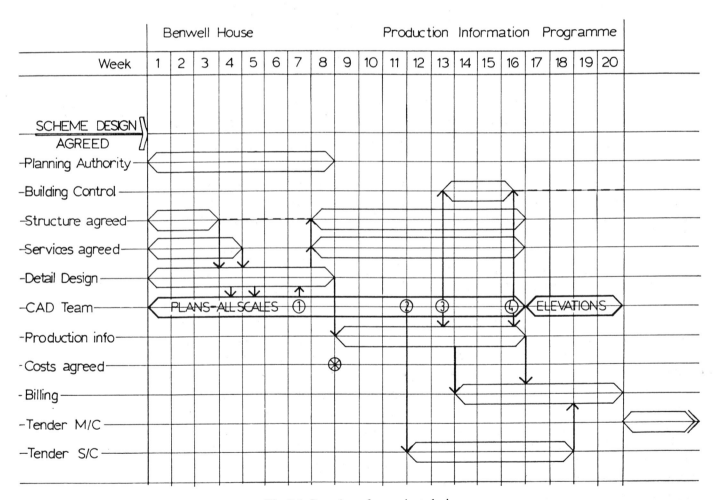

Fig 8.1 Bar chart for project design.

Project : **Benwell House**

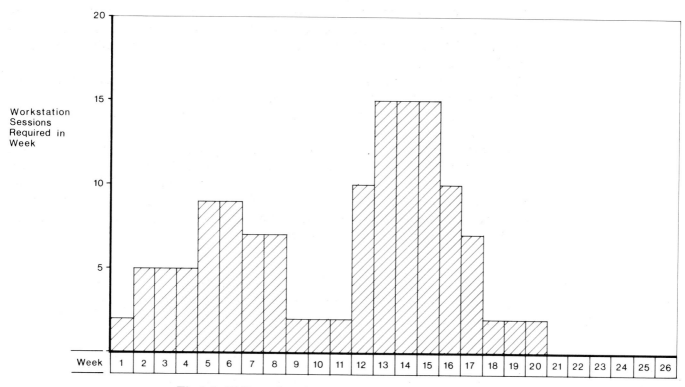

Fig 8.2 CAD workstation sessions. Project planning chart.

For the majority of design projects it is best to construct a series of bar-charts and histograms. For example, the bar-charts can illustrate the main tasks linked with times (Fig. 8.1). A histogram can show the probable demand for resources during successive time periods (Fig. 8.2).

Histograms indicating the future demand for CAD operators and workstations are immensely valuable. Of course these might be rather tentative initially. They may be subject to repeated updating as the design solution evolves. A plan is always better than no plan.

Such charts and histograms could be produced with the aid of a computer drafting system. However there are many software packages available. Some of these run on microcomputers, and have been specially developed for business and scientific charting purposes.

Graphical output helps to condense the information and this all allows the manager to assimulate it more rapidly and accurately. Obviously the charts should be self-explanatory so as to be clear to all to whom copies will be circulated. Frequent revision of the plan may be inevitable. It is essential that the changes are communicated and indeed highlighted for special attention. This means that the charts and histograms must be regenerated quickly and easily, and circulated. A computer can make such regeneration less painful and more rapid.

8.4 NETWORK PLANNING

Planning with the aid of *network* or *critical path* planning methods is commonly adopted in construction site management. These methods are however less used in the design phase. Nevertheless there could be merit in using them for design, especially on large projects where fee competition is strong, or in any situation where a high level of design management would pay dividends.

Networking involves separating the project into its component parts or activities. The duration of each activity must be estimated. The precise order and relationship of each activity within the project must be established. This is to specify the interdependencies among activities.

The 'arrow' network method uses an arrow to represent each activity. This is the easiest for beginners to understand. In the 'precedence' method, activities are represented by boxes or nodes in the diagram. This method is generally preferred for complex networks and by experienced planners.

In the planning of design activities, the main benefits often arise just from working with a network on paper. This discipline can be very helpful in clarifying the logic of the process. It can show up more clearly the order in which various tasks have to be tackled. Likewise it highlights the timing of information exchanges and critical decisions between members of the design team or client. Potential bottlenecks become more obvious. It is possible to break off the planning work at this stage and construct bar charts to show the activities more clearly. Histograms can be constructed to show the likely demand for CAD resources.

Alternatively, the network analysis can be undertaken by well-established methods. A time analysis can be done to calculate the probable duration of the whole design project. This time analysis will also indicate which activities are 'critical' in the sense that they directly control this overall duration. Other activities will show up with a varying degree of 'float' in their timing.

Network methods can be undertaken on computers, and several programs are available for microcomputers and larger machines. The former will be adequate for most design management. The main advantage of the computer is thought by some to be this ability to rapidly calculate the start and finish times for all the activities. The computer certainly can do these calculations. However, except on very large networks, a time analysis is a fairly easy task to undertake manually. The real value of the computer is its ability:

(1) to save a network,
(2) to allow a saved network to be modified when the circumstances inevitably change in the future; then to quickly recalculate the network with the minimum of fuss,
(3) to produce high quality graphics output in the form of network diagrams, bar-charts and histograms. Likewise a variety of schedules can be printed,
(4) to undertake a resource analysis (including cost analysis) if required.

8.5 MULTI-PROJECT PLANNING

Attention has been focussed on the management of a single project. The greater problem in most offices however is multi-project planning. Here a variety of small and large jobs compete for resources. Decisions have to be made regarding how the available resources are to be assigned. With CAD management, this is indeed the main problem because of course the operators, workstations, processing capacity, storage space and

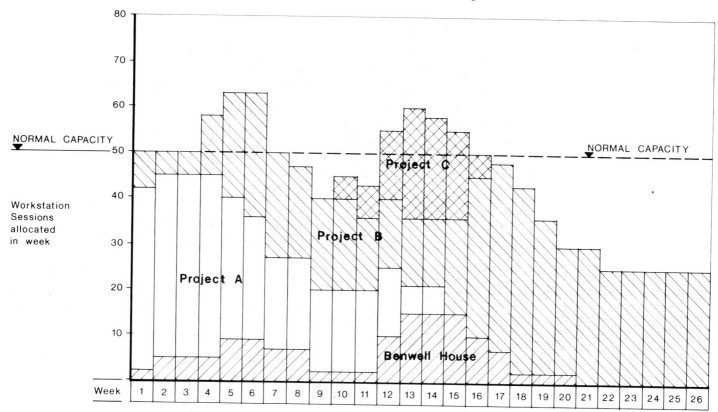

Fig 8.3 CAD system planning chart for workstation sessions.

plotter capacity are all in limited supply. They all have to be shared among several jobs and design teams.

Multi-project planning is tackled by consolidating the needs of individual projects. If histograms have been drawn for each project, showing for example the workstation sessions required in successive weeks, then it is a simple matter to consolidate these. The result would be a single histogram to show the total demands on the CAD system (Fig. 8.3). When this is done initially, the variation in demand is likely to be reflected in unacceptable peaks and troughs. The demand must be smoothed to reflect the resources actually available. This must be done by examining the individual projects again and by rescheduling some of the work. The level of resources might be modified too, perhaps by modifying shifts, by altering someone's holiday period, by the timely training of more operators or by installing another workstation.

Where fairly large numbers of jobs are involved, much useful planning can be done with the aid of either a spreadsheet program or a database management system. Preferably these should be linkable with a business charting program.

Where networking has been used, multi-project planning can be tackled by simply extending the network. The process is made easier if use is made of a program that can handle sub-networks.

In the CAD context, the main responsibility for each project must still lie with the project manager or chief designer. Where there are several projects, clearly the CAD co-ordinator must play the co-ordinating role which the title of the appointment suggests. The allocation of workstation time and trained operators on the basis of project priorities is both important and difficult in practice. On a strategic basis it can be worked out with histograms or other planning methods. The forward planning of individual workstation sessions must of course be done with the rotas discussed in Chapter 6.

8.6 SHOULD CAD BE USED FOR THE PROJECT?

One of the major responsibilities of the CAD co-ordinator is to advise the project architects or engineers on the application, capabilities and availability of the CAD system.

The system capacity is usually not sufficient to cope with all the design work of the office. In any case, some projects or parts of projects might be done better on the drawing board. So in conjunction with the project professionals, the CAD co-ordinator must continually scan the incoming projects and on-going work. Using subjective judgement, the best projects – or parts – for CAD application must be picked out. The criterion must be to use the CAD system for project work where the potential benefits appear to be greatest. It is a matter of gaining the best possible return on the investment in CAD.

As experience is gained, the CAD co-ordinator will steadily become more adept at judging the most suitable work. The following can be used as a guide to the things to look for in projects:

1   *Repetition of projects or buildings*

This is where one project is being, or is likely to be, repeated on another site. Of course, if no significant changes to the design are envisaged then quite simply, copies of the original drawings could be made. CAD is most beneficial, however, where some design amendments are necessary, perhaps to suit the new location. The copy and adapt capability to be discussed in Section 10.4 applies. A building that needs to be handed, presents no problems.

2   *Repetition of zones or floors, and symmetry*

It is common for wings of a building complex to be duplicated. Different floors in a multi-storey building can be similar to one another. Again exact replicas are simple to deal with by manual methods. It is when adaptations are necessary that the CAD method scores. Smaller zones like lift/staircase cores and toilet areas are often repeated. Many other examples, like bedrooms in a hotel, will be discussed in Section 10.4.

Where there are lines of symmetry, one half only can be designed. A copy is made of this, it is handed (or reflected) with the CAD system, and this is joined to the first part.

3   *Modular construction with much repetition of components*

This relates to projects where there is much repetition of discrete components, elements, symbols, other graphical items or standard blocks of text. If many of these components are already held within the CAD system, this would be very favourable. A CAD component library relating to manufacturers' components would be an asset.

Placement of such components on a modular grid is a task which can be quickly and easily achieved with a modelling system.

4   *Multi-discipline design – especially for general arrangements*

In many projects, the general arrangement drawings contain much background information that is repeated and reused, albeit at different scales. The changing scales is no problem to a CAD system. These drawing sheets are of course produced for different purposes, or to show the work for the various building trades. They might be produced by different design disciplines. A few examples of often-repeated graphical material found on general arrangement are:

- grids,
- building outlines,
- structural outlines,
- key plans to be included on sheets.

The general arrangement sheets with such information repeated on them include:
architectural layout drawings,
fittings and finishes drawings,
room numbering plans,
structural outlines and steel framing drawings,
reinforced concrete and structural detail drawings,
plumbing services drawings,
heating and cooling service drawings,
ventilation drawings,
controls installation drawings,
sprinkler system drawings,
reflected ceiling plans,
room numbering plans,
furniture layout plans.

In Chapter 10, we shall discuss how such drawings can be produced with the intelligent use of classification schemes for CAD components.

Conditions for CAD use are best in projects where the designer is able to obtain good information from the various sources, including the client, when required. Favourable conditions are also when there is a willingness by all members of the design team to contribute positively to the adoption of CAD.

5 *Co-ordination with elevations, section views and perspectives*

Co-ordinated plans, elevations, section views and 3–D perspectives can be produced with suitable CAD software, provided adequate graphical/geometrical information has been built up within the model. As discussed in Sections 10.3 and 10.4, this ability to co-ordinate is especially useful in complex jobs or in critical parts of buildings.

Co-ordination of the building services can be helped using CAD methods. This is design work which may qualify for additional fees.

6 *Complex geometry*

CAD is useful where difficult geometrical layout problems have to be resolved and where a precise dimensional control must be maintained throughout a project.

7 *Prototyping at preliminary design stage*

As discussed in Chapter 7, large or prestige projects can benefit from model-building during the preliminary design stage. Many perspective views of the exterior or interior of a model can be created easily as a by-product after the model has been assembled. There are marketing issues involved here as well as technical ones.

8 *Probability of model reuse for design revisions*

CAD methods can generally cope well with projects where the drawings are likely to be exchanged frequently among design team members. Drawings often are worked up gradually from scheme drawings to become detail design drawings, thereafter perhaps to become tender drawings, construction drawings and eventually to form the 'as-built' records.

Large projects with several design disciplines working together, and where many outside authorities are involved, are the projects which will have the most revisions to drawings. CAD can handle many revisions fairly easily. It might take just as long to revise the model and extract new drawings as it would to make revisions to manually produced drawings. However, the CAD system produces new and clean drawings every time. This is more than can be said for the manual method. Likewise, the drawings from the CAD model are more likely to contain consistent information.

9   *Standard details*

Tried-and-tested standard details are much talked about but not so often actually used on projects. In traditional design, attempts to reuse details often fail because the retrieved details always seem to require modifying so as to suit the new circumstances.

With CAD methods in use, the saved 'standards' can be readily available at least as a source of ideas and some graphics. The copy and adapt facility of the CAD system should be able to cope well. When many archived details look as if they could be suitable for the new project, then conditions for CAD use are certainly favourable.

It is not necessary to apply CAD methods throughout a design project. For example a decision might be made not to use it for the preliminary design. Alternatively a decision may be made to use it for the general layout but not for design details to be shown at the scale of 1:10 or larger. Perhaps one building in a project might be designed using CAD while another is tackled by manual methods.

When CAD and manual methods are both to be adopted on the same project, then a rigid and logically-based division must be applied. It must be adhered to as well. In no circumstances should computer and manual methods be mixed for producing the same sort of design documentation within one project. Thus we should never have some of the 1:50 architectural layout drawings done by hand, while others containing overlapping information are done by CAD. Failure to adhere to this will limit the benefits of CAD. It will also lead to much confusion.

Worse still is the situation where a drawing is produced with the CAD system. Afterwards a design revision is necessary and the alteration is noted manually only on the plot. Perhaps it was not easy to get to a workstation at the relevant time. The design model held within the computer will now be out-of-step with the drawing, and this will certainly lead to future problems. In these circumstances the plot must be altered and later the model must be changed as well.

8.7 AGREEMENTS AMONG DESIGN TEAM FOR A PROJECT

Before much work is embarked upon, certain agreements must be reached among the members of the design team and the other parties involved. Some of the following points are important whether or not CAD is in use – sadly the quality of project management in design is so poor that these basic agreements are rarely even considered.

● The key dates for information exchanges. These will usually be important milestones to be included within the project plan. For example, an architect may release floor layout information which becomes the background on which the building services engineer has to lay out his work.

Usually these information exchanges will be drawings and schedules in hardcopy form, i.e. on paper. The information requirements of each organisation must be determined so that the contents of drawings and schedules can be planned and the documentation produced with the CAD system. This was discussed in Section 8.2 under Detail Design.

- Early agreement is necessary on matters like drawing sheet sizes, drawing frames, title blocks, titling, drawing numbering, drawing register formats, scales to be adopted, and other drawing standards like dimensioning arrangements and symbology.

- Sometimes CAD files containing drawing information or design models will be exchanged in machine-readable form, such as on magnetic tape or discs. In this situation, it is advisable to agree on a classification scheme for the CAD data. This relates to the construction elements, trades and so on.

   Likewise, arrangements will have to be made regarding the data transfer procedure, the media to be used and data formats. Usually a trial exchange should be carried out as discussed in Section 6.10. This will ensure a smoother transfer of the real data.

   It should be borne in mind that different firms are usually at different stages of computerisation. Commonly accepted procedure in one firm may be unknown in another.

- Besides the pre-arranged issues of design information, offices have to issue design revisions from time to time during the currency of the project. A procedure must be agreed among the parties for controlling the issue of revised drawings, schedules or computer data. I will deal with this later in Chapter 12 under the heading 'Revision Control'.

CHECKLIST FOR PROJECT MANAGEMENT

1   Recognise the need for project management.

2   Should CAD be used for the particular project?

3   Will CAD be used for the preliminary design? If so:
   (a)   define objectives,
   (b)   construct programme for preliminary design work,
   (c)   fix dates for modelling work, choose operators and assess requirements for system work-sessions.

4   Will CAD be used for the detailed design? If so:
   (a)   list information required from others – with target dates and names,
   (b)   list problems/key decisions – with target dates and names,
   (c)   list drawings/schedules to be produced in-house – with target dates.

5   Plan project design operations, using:
   (a)   bar charts, *and/or*
   (b)   network planning, resources histograms.

6   Determine CAD resources required for detailed design:
   (a)   operators,
   (b)   workstation sessions – histogram,
   (c)   design information available from other projects.

7   Multi-project planning.
   (a)   Consolidate plans for individual projects.
   (b)   For CAD, consolidate histograms for workstation sessions.

8   Agreements among design team for a project.
  (a)   Key dates for information exchanges.
  (b)   Drawing formats and conventions.
  (c)   Will CAD data exchanges be made? If so:
    - agree design information classification schemes,
    - define data transfer procedure,
    - undertake a trial transfer.
  (d)   Agree on procedures for document issue and revision control.

CHAPTER 9

Components: Graphical Representation

Just as real construction projects are built from bought-in components and elements constructed on site, so computer models are built from CAD components. This chapter deals with the graphical definition of various kinds of these. Apart from fully pre-defined components, the concepts of component hierarchies and of variable components are explained. The latter can have one or several dimensions left undefined at the time when the component is created, and this functions as a shorthand method of dealing with families of similar – but not identical – items. The programmable component is a more sophisticated concept, and this allows for an element of automatic design or selection of components for particular circumstances.

The management strategies for creating and saving components are discussed, as are the setting up and use of libraries. The master library of CAD components soon becomes a valuable asset for the design office.

In Chapter 10 we shall see how large numbers of components can be assembled to form a digital model of a real project.

9.1 INTRODUCTION

A draftsperson produces a drawing manually by drawing the lines and annotation on the drawing sheet. Any 'structure' incorporated into the drawing is implicit. It lies in the mind of the draftsperson and must be interpreted by the users of the drawing.

With CAD, the lines and text can be composed into a drawing, but in addition a structure can be imposed explicitly on to the design information. Essentially we can create components using drawn lines and text. These components can be arranged in hierarchical form to produce assemblies, design models and projects.

Components therefore lie at the heart of most CAD systems. Most of the CAD operator's time and effort is involved in the input of graphical data. Naturally everything possible must be done to reduce this input burden. One means is to make full use of components which recur many times. Each item is defined once, and thereafter we can call on the computer to replicate the information that it already contains.

Components can be represented by graphical or geometrical information. Indeed this is precisely what many CAD systems concentrate on, for this information eventually allows the drawings to be produced from models. First of all, therefore, we shall explore the creation and storage of the graphical representation of CAD components.

Design information is not limited to graphical data only (Fig. 9.1). The usefulness of a CAD system is considerably enhanced if components can be represented also with non-graphical descriptions or attributes. To provide these facilities, ideally a CAD

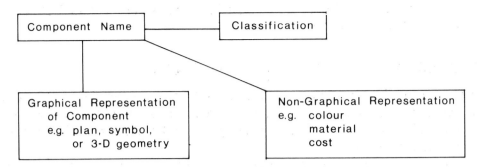

Fig 9.1 Representation of CAD components.

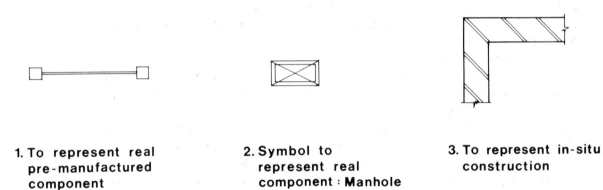

1. To represent real pre-manufactured component

2. Symbol to represent real component : Manhole

3. To represent in-situ construction

4. Standard Detail

5. Graphics Object

Fig 9.2 Component types.

system should include or be linkable with a database management system (DBMS). This is a subject which will be tackled in Chapter 11.

9.2 THE USE OF GRAPHICAL COMPONENTS

The basis of many CAD systems is that it is possible to create any number of graphical objects, elements or components. These can be saved in the computer for reuse later in the assembly of design models. The terminology differs but throughout I intend to use the term *CAD component*.

The creation and assembly of CAD components mirrors to some extent the construction of a real building from bricks, windows, sanitary fittings, and many other items. However CAD components do not only relate to the real building components. Indeed CAD components can be any of the following items (see also Fig. 9.2):

1   Diagrams which represent the outlines and other details of the real pre-manufactured components brought to site and used for construction.
   Examples: cladding panels; doors; windows; hot-water radiators; air-conditioning ducts; pre-cast floor beams; roof trusses.
2   Symbolic representations of the real pre-manufactured components.
   Examples: symbols for light switches to appear on floor plans; manhole symbols for small-scale drawings; special line-styles used to represent the lengths of water pipe or cable runs on drawings.

   Symbols are not intended to mimic the real appearance of components. They are generally accepted as conventions or standards, and no modification is made to a symbol when it is being included in a model or drawing.
3   Means of representing the outlines of *in situ* or monolithic construction, such as brickwork walls, concrete floor slabs, asphalt roofing.

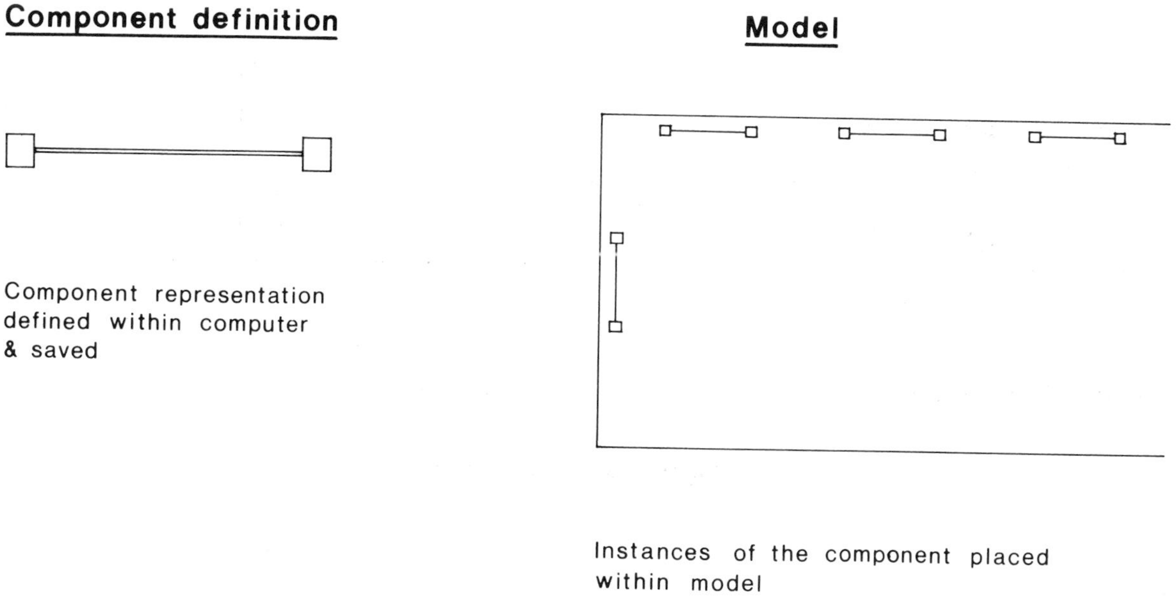

Component definition   Model

Component representation
defined within computer
& saved

Instances of the component placed
within model

Fig 9.3 Component definition and instances.

4 Standard Details. Usually these are diagrams which are composed for inclusion on drawing sheets. Usually they explain how some part of the construction is to be carried out.
   Examples: typical manhole construction; expansion joint in a floor slab; damp-proof course arrangement; or a typical parapet construction.
5 Recurring graphics objects or blocks of text required for drawing clarity or to assist the construction process.
   Examples: grid lines; dimensions lines; north points; drawing frame; title box; blocks of standard notes.

These are all examples of recurring CAD components used for producing design models and drawings. The creation and use of components works well in practice because the computer is of course an ideal tool for storing, retrieving and manipulating all such information. Each component is defined and saved once only. Each time the same component is required for placement in a design model, we actually place an 'instance' of the original component (see Fig. 9.3).

9.3 PREDEFINED COMPONENTS

Component name and classification

When a new component is to be created, first a suitable name must be assigned to it. This must be an unique code name because this is how it will be retrieved or reference made to it later.

Large numbers of different components will be assembled into a building design. It is essential to be able to classify them, because we must have some method by which we can select certain classes of components within the model. Such selection will be necessary for:

- displaying design information on a screen,
- plotting of drawing sheets,
- printing of schedules,
- when design information is to be copied or deleted within projects.

Many drafting systems adopt a system of drawing layers for classification purposes. In the Sonata and Rucaps modelling systems for example, one of the means of classification is by assigning each new component to a specific 'Category Number'. This then becomes an inherent property of the component. For example all doors might be assigned to one category, windows to another, internal partitions to a third, external walls to a fourth, ventilation ducts to a fifth, furniture to a sixth, and so on. With an adequate range of categories, the classification by category can be as detailed as required. Thus we could assign concrete columns, steel columns and brick columns to three different categories.

Classification of components provides powerful manipulative facilities to the designer. Also, it is by such classification that the database within the computer can be given a structure. This is of such basic importance that the user must be prepared to give much thought to this component classification.

It would be possible to allocate the classes of information to suit each project. In practice it is far better to produce a basic system of classification for the organisation as a whole, to be used on every project. To cater for the peculiarities of individual jobs, it is always possible to allocate some of the 'vacant' classes as required. The CAD co-ordinator must have the overall responsibility for producing such a basic classification scheme for the whole design office. Clearly this should be done early in the CAD implementation phase. The CI/SfB system is a classification scheme which is widely used in the construction industries within several European countries. This, or another suitable scheme might form a basis for assigning component classes.

The concept of classification of design information is being introduced here, however we will return to this important subject in Section 10.5.

Table 9.1 Component representation in various types of CAD system

Model type	Available component representations
A 2–D drafting	Single 2–D diagram or symbol
B 2–D modelling	Single 2–D diagram or symbol
C 3–D locational model with 2–D component images	One or several 2–D diagrams or symbols e.g. plan, elevation and section of component
D Wireframe model	Edges of component
E Surface model	Surfaces and edges of component
F Solid model	Solid entities, surfaces and edges of component
G 3–D locational model with 3–D component representations	3–D definition of component
H 3–D locational model with 2–D/3–D component representations	One or several 2–D diagrams/symbols and/or a 3–D definition

Component graphics

The means of representing objects or components is summarised in Table 9.1. It is something which differs in the various types of CAD system.

In drafting and 2–D modelling systems, the graphical representation of each component has to be a single 2–D diagram. It may be a plan view of the component outline, or an appropriate symbol by which it can be recognised on drawings.

In a 3–D locational model which uses 2–D component images – the Rucaps system is an example – each component can be represented by either one or several 2–D images. We saw in Chapter 3 that a window in a building, might for example be represented by three diagrams, namely a front elevation, a sectional plan view, and a sectional end elevation (see Fig. 3.11). When drawings are being extracted from the model – whether they were plans, elevations or sections of the building – the computer would pick the most appropriate of the available images to represent the component.

The component might be defined in a 3–D system in terms of its edges, its surfaces and/or solid entities, depending on the proprietary CAD system being used. The definition may include the inherent colour or transparency of each surface. When such a definition is available then the component will show up in 3–D views which are automatically generated. If surface colours are defined then the component could appear in colour shaded views of the assembled building.

Some CAD systems adopting the locational model concept permit any component to be represented either with 3–D geometry and/or one or more 2–D images such as t²'s Sonata system (Fig. 9.4). The modelling software must be able to gain access to the defined component representations when required. When more than one representation is provided, the computer will select the most appropriate for the task in hand. This provides much flexibility to the designer. For example he would define in 3–D only those components which he wants to appear on 3–D drawings. Other components would be defined in 2–D only, thereby gaining the benefit of relative simplicity and reduced volume

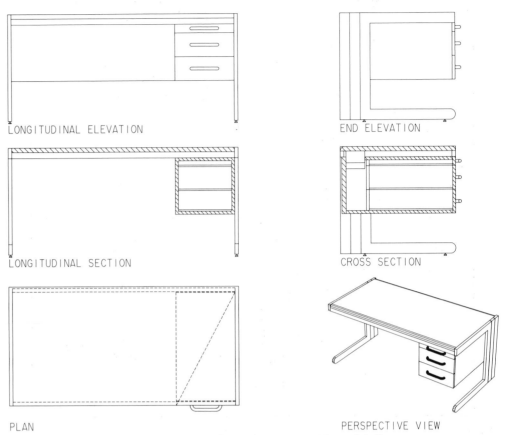

LONGITUDINAL ELEVATION

END ELEVATION

LONGITUDINAL SECTION

CROSS SECTION

PLAN

PERSPECTIVE VIEW

Fig 9.4 Component representation in 2–D and 3–D.

of input. Adequate 2–D information could be provided for all those components that will be required to appear on co-ordinated plans, elevations and sections of the project.

The designer must think carefully when components are being created about their subsequent use. An eye must therefore be kept on the nature of the design documentation that will eventually be required.

Fig. 9.5 shows a component data sheet and the 2–D and 3–D representations for a handwheel-operated ball valve as retained within PDMS – Plant Design Modelling System. The component clash detection volume is a user-defined 3–D envelope into which other components in the plant model must not intrude, thus ensuring ready access to the component in the real plant for operation and maintenance.

Components are always defined by inputting their actual sizes, usually in millimeters. With full-size information always stored, the computer is well able to re-scale the graphics when any suitable drawing scale is specified later. Of course some special thought is required when a symbol is defined.

When text annotation is to be incorporated with the graphics, the actual height of the plotted text must be quoted. It is best to make the text height to suit the most commonly used drawing scale. This could be the general arrangement scale of 1:100. At other drawing scales, the text might appear bigger or smaller depending on the system in use. It is always preferable if users can control the text height independently from the graphics.

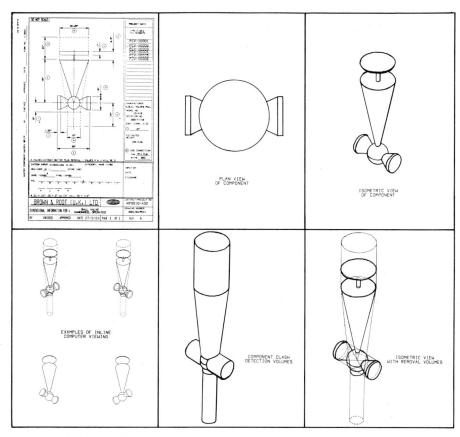

Fig 9.5 Component definition for use in plant design. (*Courtesy*: Brown & Root, Vickers. London.)

9.4 COMPONENT HIERARCHIES

The concept of a hierarchical set of components being grouped together or assembled to form a higher-level component is quite common. For example a table, six chairs and a sideboard might be grouped to form a suite of dining-room furniture. An office desk, chair, filing cabinet, bookcase and partition might constitute an office-workstation. A frame, door leaf and door furniture together form a door-set.

Such component hierarchies prove useful in CAD work. For example we might wish to position several copies of a complete office-workstation. Alternatively we often want to move or otherwise manipulate *en masse* all the components in the hierarchical set.

A hierarchy may have just two distinct levels, although ideally we need multiple levels. Then, for example, a door lock can belong to the door furniture, that belongs to a door set, that can belong to the entrance foyer of a building.

If it is possible to group components into an assembly and give it a single name for manipulation purposes, it must also be possible to 'burst' the assembly back into its constituents. Such facilities require some thought before embarking on their widespread use, but they certainly help to add flexibility and structure to the design model.

9.5 VARIABLE COMPONENTS

So far, I have discussed ordinary components which have their full-size dimensions defined at the time of their creation. All that remains to be done is the placement of instances of each component as required within a building design.

When dealing with a wide range of discrete components, this is a straightforward and simple procedure. It is the appropriate procedure to use when we are dealing with a component like a washhand basin which, as far as construction drawings are concerned, need not vary in size or shape.

Let us now look at a pre-cast concrete floor joist. In certain buildings, all the joists would have the same length. A pre-defined component would be created to represent this unit, and then many identical instances of this would be placed as necessary. However another design might incorporate a multitude of joists, each instance having a differing length. Here we would have to laboriously define many joist components, one for each differing length required. To avoid this needless labour, the concept of the variable component has been introduced. These are sometimes referred to as parametric components, and other terms are sometimes used.

Variable components are those which can be initially defined in such a way that one or more of their dimensions or parameters can be treated by the computer as variables. All their other dimensions and properties are fixed. In this way, perhaps a single component definition can be saved in the computer to be available to represent a wide range of components which are basically similar – but not identical.

Of course all the dimensions without any exception must be defined before any drawings containing the components can be displayed or plotted. So the variables are in practice specified or fixed only when each instance of the component is placed within a design. It is really a matter of convenience. For ranges of similar components, all the dimensions and properties which do not vary are specified once only. The variables have to be specified for each use. Several CAD systems provide facilities in one manner or another for variable components.

Some components tend to have a well-defined cross-section yet within a design they are used in differing lengths. An example is a universal-beam section used in structural-steelwork, such as the $305 \times 102 \times 25$ kg UB. This section might recur many times as members of differing lengths within a structural frame or roof truss.

In the original component definition, all the dimensions of the component, except one – its length – are stored in the computer. Later, each time the component is retrieved for assembly into the building model, a missing length must be supplied by the operator to complete the definition of an instance. The operator might merely point to the locations in the model of the beginning and end of the instance, and the system then calculates the length.

Variable-length components are useful for representing partition walls, concrete columns, and air-condition ducts where the cross-sections have been specified.

Where several variable dimensions are permitted, this can be a shorthand way of defining a whole family of similar components. As an example, a single component could represent any straight piece of pipe. Two variables could be arranged to represent the pipe nominal diameter and the length. Another example is that a single component could be defined to represent any straight piece of rectangular-section ventilation duct. Three variables would be width, depth and length. A rather similar example would be a component for any rectangular concrete column. Other multi-dimensional components could be devised to represent doors, windows, radiators and many other familiar building components which are supplied in a variety of different sizes.

Programmable components

The variable nature of components can be extended either by use of programming languages provided in systems for CAD users, or by special facilities like the Sonata Parametric Elements. An insight into what might be done using such facilities follows:

- Several variables and/or text strings can be defined for each component. Fixed values need to be supplied only when an instance of the component is placed at building assembly time.
- Default values can be allocated to the variables when the component is created. When each instance is used, the operator can either leave a default value intact or he can substitute any other value.
- A component can be defined in such a way that at assembly time, each variable can be automatically queried. This can be done by the display of a prompt which can be phrased in terms that the operator will readily understand.
- Further derived values in the component's definition can be calculated from the variables. These calculations use equations supplied when the component is originally defined. Calculated values could form a part of the text which is associated with the component. This feature allows dimensioning annotation to be set up.
- Checks, comparisons or calculations can be included within the specification of the programmable component to ensure that the user-supplied values are within reasonable bounds. If any are not, the user either could be informed or else suitable values could be imposed.

A single programmable component could represent a complex family of building components. Examples of these are:

- a set of concrete retaining walls,
- a range of window types,
- all oval ductwork reducers,
- all available universal steel beam sections,
- a set of rectangular pitched roofs.

The inbuilt equations can represent design rules. These might be either designer's 'rules-of-thumb' or scientifically-based design calculations. They could also represent the requirements of official standards, codes of practice or the national building regulations.

Let us look at two examples of the application of programmable components:

1 A programmable component could be defined to represent all the timber joists in an area of floor. When joists are required for a room, the operator would invoke the component, and specify values for the span, floor-loading and the room width. The results in graphical form, suitably annotated, would be reported to the user. The group of joists could be placed where required in the design solution.

   The computer would determine the size, number and spacing of the joists. The arrangement would conform with input values (the variables). The equations inbuilt to the programmable component might be based on two design rules-of-thumb such as:

   (a) the depth of a joist is the span divided by 24 + 50 mm, where depth and span are in millimetres;
   (b) the width of the joist is one fifth of the depth.

   Further equations would round the calculated values to commercially-available timber sizes.

   Clearly this arrangement of joists might be suitable as a preliminary design. It might have to be checked later by more rigorous methods before being formally adopted.

2 The complete geometry of a staircase conforming to the building regulations could be calculated automatically from a pre-defined programmable component. This

geometry includes the dimensions of risers and treads, the number of risers, the width of the flight and the height of the handrail.

As with any other type of component, one or several 2–D images could be defined with variable dimensions so as to represent the programmable component. The computer can then calculate how the instances should appear on plans, elevations or section views of the building. It is likewise possible for the computer to automatically set up the 3–D geometry. Of course this is pre-defined in every way except for the variable dimensions.

Once created, an instance of the programmable component can then be invoked and assembled within any building design. This instance then behaves in a similar way in the CAD system to all other components. Thus it could be placed, copied, moved and revised as the operator wishes.

Variable components provide more flexibility

We can now see that any variable component – and particularly the programmable type – is a powerful facility within CAD modelling. After it has been defined and saved, it may be retrieved later for use. Then it can be stretched, shrunk, distorted or even may be specially designed to meet any local needs.

We must not be tempted to think that design options will be restricted because models are assembled from pre-conceived components. In reality the situation is highly fluid. With a little creativity, the designer is not likely to suffer from any limitations being imposed on his expression.

9.6 STRATEGY FOR CREATING AND SAVING COMPONENTS

When a component is created or selected, the user has to keep in mind the intended method of its assembly into the project model, as well as the drawing output that will be required. Only by doing this can he decide how the design solution should be represented in the computer model.

Many feel that a huge library of components must be created first, before any real design work can be done using a CAD system. This is not so. Certainly it is useful to have a few components available initially, and indeed some can be produced by the operators during their initial training sessions. The CAD co-ordinator ought to give some prior thought to this matter so that such training time is used as productively as possible.

For the rest, it is probably better to create components during the progress of the design work. This way little time is wasted and truly practical components are more likely to emerge. Project components are mostly created on the initiative of an operator or project designer. Each operator soon acquires his own set of tricks-of-the-trade, for example for using CAD components to model the monolithic building elements like slab floors and brick walls.

Nevertheless component creation is something which should be planned and co-ordinated. So the CAD co-ordinator has a central role here. Ideally this person would be an experienced designer, have a good knowledge of the CAD facilities, plus some knowledge of both existing and imminent project work. If at the same time the CAD co-ordinator actively communicates with the operators then duplication of effort can be avoided.

Together the CAD co-ordinator and operators must work out how the design can be represented using components of the various types. If someone creates a component or design features which are potentially useful elsewhere, this fact should be spotted by the co-ordinator.

The operating skill level is likely to vary within the organisation, and indeed such variation should be encouraged as an asset. Those with low levels of CAD skill may be able only to use the components already created and made available by others. The average operator should be able to create ordinary components, and perhaps variable-dimension ones as well. Programmable components are an advanced concept and perhaps only one or two of the most skilled and enthusiastic operators will work at this level. The fruits of their activity – pre-defined components – must however be readily available to others. The CAD co-ordinator must ensure that the most useful components are not hoarded purely for personal or sectional advantage.

9.7 COMPONENT LIBRARIES

Components which are potentially useful elsewhere are stored in libraries. A library must be easy to create, access and maintain. Also it must be compact and must allow for growth and modifications. Perhaps it is now clear that a library consisting of a number of ordinary components, plus a modest number of well-chosen variable components, can soon become a valuable asset within the design office.

In practice it is possible to recognise component information at three levels:

1 *Externally-sourced CAD components*

These might be standard symbols recommended by the national standards authority or the products of firms that supply construction components.

Some CAD components might be available directly from the CAD system vendor or from other design offices that are using the same proprietary CAD system.

Two sources of CAD components in the UK are:
- RIBA Data Services,
- Architectural Data Services Ltd.

Each has created an extensive library of the component ranges of several of the leading suppliers of building products. This data is available to user firms on magnetic media in formats to suit several of the main CAD systems.

2 *Company master library of CAD components*

These are components created in-house. They may have been created specially for this library, but more likely they will have been transferred from various design projects. The criterion for retaining them in the master library is merely that there is some potential for their use on future projects. See Section 9.8 for further discussion.

3 *Project library for CAD components*

This is where all the components which are required within a current project model are kept. This will include all the doors, windows, walls, and so on.

Here some components will be created specifically for the project, some will be taken from the company master library, and others may be taken from external libraries. 'Taking' components does not mean that they are actually removed from the source. It is often merely a copying process, analogous to photocopying a document. Sometimes however it might be a 'copy and adaption' process which starts with some pre-existing material and then modifies this to become something that will be more suitable for the particular circumstances of the project. This can be much quicker than creating a new component from basics.

9.8 THE COMPANY MASTER LIBRARY OF CAD COMPONENTS

The maintenance of this master library is the responsibility of the CAD co-ordinator, even if much of the work is delegated. Components used for one project need to be scanned at an appropriate time. A copy of everything that appears likely to be useful elsewhere must be acquired by the CAD co-ordinator and copied into the company master library. Items must of course be carefully checked before being incorporated. Likewise the CAD co-ordinator must ensure that operators are not re-creating components that are already available.

The master library is a vital element of the company's CAD strategy. The initial objective should be to include the most commonly used design features and components. After a time the master library becomes the immensely valuable repository where project building-blocks are kept. It will soon become the storehouse which contains in tangible (i.e. machine-readable) form much of the experience of company. A well-stocked and well-co-ordinated master library grows in value as time passes. As it does so, the productivity of the CAD users grows with it, and design quality is enhanced.

Components must be grouped using an easily understood naming scheme. They should be stored permanently on disc so as to be readily accessible to operators. Operators and designers must be provided with a pictorial catalogue of the master components. This will be in A4- or A3-size format and kept in a suitable ring-binder. This must be updated frequently.

In the master library, the distinction needs to be made between components containing geometrical description, and symbols.

The former have full-size dimensions and can be automatically scaled when drawings are produced. Also they may be edited before use.

Symbols on the other hand are usually standardised and so are never modified. This means that relevant standards for symbols should be settled early in the system implementation phase, so that there is little likelihood that the symbols already placed will have to be changed later.

Graphical symbols, drawing frames, standard details, standard sets of notes and such items would certainly be retained in the master library when there is a desire to standardise their use throughout the firm. This indeed is the realistic way of imposing a company house-style for its documents.

Again I must stress that throughout I have been using the term 'component' in a very wide sense. It is to represent either the real building components, the building-blocks for monolithic construction, graphical symbols and text annotations, diagrams or construction details. The component might well include non-graphic attributes and descriptions as well. A library of many of such components is undeniably useful. The facility in a CAD system to pick out such geometrical components and/or details, and then to copy them to a project either as they stand – or to make a copy which is then adapted to fit particular new circumstances – is something that should not be underestimated.

Plate 3 is an excellent illustration of the use of 3–D components. The geometry of a number of objects is defined and stored. Views of single objects are ranged around the edge of the plate. The model of the building is built up by placing and orientating these objects. After this is done, the model – the building interior – can be viewed from any point. Obviously it is simple to rearrange the objects within the model, and obtain other views.

CHAPTER 10

Model Assembly and Drawings for Detailed Design of Projects

This chapter deals with the use of computers for project detailing. It is not usually a practical proposition to build a full 3–D model which contains all the details of a complete project. Yet to simply produce one drawing after another with a computer drafting system is to miss opportunities. So the discussion of CAD techniques, initiated in Chapter 3, is taken up again and extended to cover the design detailing phase. In practice, a pragmatic blend of 2–D and 3–D computer modelling appears to be the best answer.

Model building procedures depend on the proprietary CAD system that has been adopted, but some of the principles involved are examined. Of course, models are normally built up by assembling CAD components. Designers should however also look for opportunities to reuse existing models or parts of models. Some observations are made on multi-user working with computer models.

An examination of the classification of design information and design layers follows because here lies the essential key to the manipulation of design documentation. This leads on to the discussion of how drawings can be extracted from computer models, and how sheets are composed and plotted. Different CAD systems adopt different practices here.

There is a brief discussion on drawing standards and conventions. The chapter ends with suggestions for the referencing of computer models and drawings.

10.1 REVIEW OF CAD TECHNIQUES

The assembly of a detailed design model can be handled by one or more operators using their graphics screens and input devices. The main CAD techniques were covered in Chapter 3. I ended there by stressing the importance of adopting a modelling approach to suit the application in hand. Let us remind ourselves of the various approaches discussed there:

A. *2–D Computer Drafting*
   - which mimics the drawing-board method. One drawing is produced after another.
B. *2–D Modelling*
   - where 2–D images represent the components or elements of construction and these are assembled into a 2–D model of a project. Drawings can be extracted with specific classes of design information or for specific areas of the project. In practice separate sub-models are needed for plans of different levels, for elevations and for each section through the project.

157

C. *3–D Locational Model with 2–D Component Images*
- 2–D images of components are placed or located in 3–D space. Plan drawings can be extracted automatically for any area, for any level. Where a component can be represented by several 2–D images, then it is possible to extract co-ordinated plans, elevations and sections. These drawings should all be consistent with each other.

D. *Wireframe Model*
- 3–D representation with the edges of objects defined. Line drawings, including perspectives, are possible.

E. *Surface Model*
- 3–D representation with edges and surfaces defined. There is potential for high quality visualisation.

F. *Solid Model*
- 3–D representation with edges, surfaces and interior of objects defined. This is a fuller description of objects, which allows measurement and clash detection. There are practical limits on model complexity or size.

G. *Locational Model with 3–D Component Representations*
- Individual components are defined in some form of 3–D representation. Instances of components are placed or located in 3–D space to form assemblies. This model is useful for assemblies like buildings where some components tend to recur several times.

H. *Locational Model with 2–D and/or 3–D Component Representations*
- Individual components can be defined with 2–D images and/or 3–D geometrical description, with the designer making the decisions. Component instances are placed in 3–D space location. This model has potential for large and complex assemblies where components types tend to recur. Drawings, including 3–D perspectives, can be extracted and are all likely to be consistent.

10.2 WHICH CAD TECHNIQUE IS SUITABLE FOR DETAIL DESIGN?

In Chapter 7, we looked at how a design project could be tackled with a CAD system. However the subject there was the preliminary phase of design. We saw how modelling techniques could help the client, designer and others to visualise and assess several schemes for a project. The amount of design information contained in such models is not likely to be excessive. Accordingly it is often practical to use some form of 3–D modelling (e.g. techniques **D** to **H** above) to obtain realistic perspective views of schemes. This is particularly useful when project aesthetics is paramount.

Given the choice, many project designers would be inclined to opt for 3–D modelling for all their design work. It seems inherently better, and indeed 3–D modelling does provide a fuller geometrical description of the design solution. However it is essential that any such enthusiasm is tempered by the practical issues.

Limitations of the 3–D definition for detailing

In this chapter, we have turned our attention to detail design and here the conditions are different than in the earlier stages. In particular, it is unlikely to be a practical or economically-viable proposition to use 3–D modelling for the detailed layout design of an entire project. The volume of design information is far greater than in the earlier design stages. The operators would find the input to be too onerous, and the computer's rate of response is likely to be inadequate. As a general rule therefore in the detailed design of whole projects, the bulk of the work has been done in 2–D, using approaches **A**, **B** or **C**, noted above.

Eastman[32] has pointed out that it is not merely a matter of waiting until adaptable and user-friendly solids modellers are available. He argues that there are other reasons why the solid modelling approach has not been universally successful for building design. The use of paper for drawing work has both forced and allowed designers to abstract the various issues associated with their designs. Various 2–D abstractions have evolved to allow these issues – like circulation, air flows, structural grids, electrical services – each to be considered in isolation. The abstractions allow the designer to focus attention on one particular design variable, to the exclusion of others. Without such focus, complex design work could not be carried out easily.

An easily understood analogy to back up Eastman's argument is found in electrical circuit design. The schematic circuit uses the abstraction of standard symbols to represent physical components, and straight lines for the electrical connections. Such abstraction helps the designer to concentrate on the logical functioning of the circuit, without compounding the problem by introducing considerations of the physical layout of the circuit. The latter can be sorted out later when the circuit schematic has been finalised.

Eastman's argument in the architectural context is that if a solid model only is built, the designer would often have to introduce too much geometrical complexity. This would cloud the design issues being examined by him at the time. Sometimes the designer needs to use diagrams, sketches, schematics and symbology to represent the complex issues with which he is dealing. But the matter is more serious for, as Eastman explains, there are occasions when the architect deliberately does not present a full geometrical description. He chooses to leave some details to be sorted out by others. Perhaps these details are added in the shop drawings produced by the constructors, or are sorted out on site by the craftsman. Examples include dimensioning of pipes and wiring, jointing, or the solutions adopted around the edge of ceiling or floor tiles.

The uses of 3–D in detailing

Eastman does not rule out some use of 3–D models in detailed design work, nor should we. Although not applicable to the whole project, there can be scope for 3–D modelling in the detailing of certain localised zones. Examples of these are:

- For the detail design of *prestige* zones, such as the entrance foyer of a building or other public area where there is going to be much interest in the aesthetics.
- Zones which will be used *repeatedly*, such as a typical work-unit within an office building or one typical bedroom unit in a hotel design. Here it is clearly important in the overall economics that much design attention is focussed on these typical units.
- *High-complexity* zones, such as a plant room or lift core. Modelling in 3–D here can help to solve spatial problems of accommodating many components in a cramped space, and help to highlight potential clashes between them.
- Zones of *dimensional complexity* where, depending on the capabilities of the system, 3–D modelling may help to sort out measurements, and the geometrical layout generally, with adequate precision.

Modelling in 3–D is increasingly being adopted for the detailed design of some industrial plants. The substantial cost of doing this can be justified by the high complexity of the design, the dimensional complexity, the need to reduce design errors, and the scope for repeated use of standard components like pipes and steel sections. Figs. 10.1(a) and 10.1(b) show examples of this work using the PDMS CAD system.

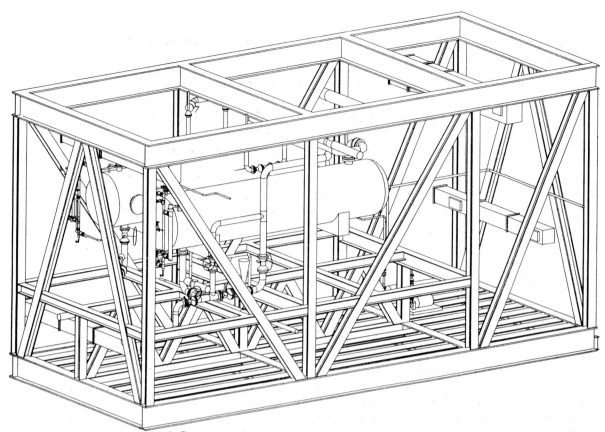

Fig 10.1(a) View of plant model. (*Courtesy*: Brown & Root, Vickers. London.)

Fig 10.1(b) View of plant model. (*Courtesy*: Brown & Root, Vickers. London.)

A blend of 2–D and 3–D

It appears therefore that 3–D modelling is part of the answer only. Eastman expresses it thus:

> 'Multiple representations break down the issues and allows them to be dealt with piecemeal; otherwise they would be overwhelming in their complexity. Thus solids modelling could never be *the* representation of design; it could only be one of several . . . The organisation of multiple abstract representations is an important research issue, not only for architecture, but for all design fields.'

Clearly there is at present no perfect CAD tool that covers all the requirements of designers. However, designers working today must choose in a pragmatic manner, the best range of tools for their current work. Table 3.1 and Section 10.1 above may provide some pointers in this choice.

For the simplest projects, perhaps a straightforward computer drafting system provides a satisfactory and easily applied tool.

A preliminary design – uncluttered with detail – can be studied visually by means of a 3–D model.

The details of large projects can be designed using a blend of 2–D and 3–D methods. Limitations of present-day hardware and software mean that the bulk of the work must be in 2–D and the minority in 3–D. As systems improve the balance may move a little, and more ambitious work will be done in 3–D. Eastman tells us that it will never all be done in 3–D. One means of blending 2–D and 3–D is of course to adopt approach **H** listed in Section 10.1, where the overall geometrical framework can be provided by the spatial locational model. The parts can be represented by symbology, schematics, 2–D outlines or full 3–D – as the designer thinks fit. Fig. 10.2 shows how the methods are blended within the plant design industry.

From the viewpoint of CAD management, we can now understand why flexibility is the key in system choice. The most useful CAD system would offer a choice between 2–D methods and some form(s) of 3–D modelling. Then it is the job of the CAD co-ordinator to try to guide project designers and operators on what is best in given situations. There is little substitute here for a build-up of CAD experience over a period.

10.3 MODEL ASSEMBLY

The components

Since CAD systems and the projects for which they are used vary so much, it is not possible to provide a detailed description of the appropriate procedure. Likewise it is not easy to lay down detailed ground-rules for how the work should be tackled. In this section, we will however discuss a few general guidelines for model preparation.

We saw in Chapter 9 that there may well be a master component library which has been built up and is maintained by the design office for the benefit of all of its projects. In addition to this, there could be one or more libraries of component information acquired from external sources.

A new component library or catalogue has to be built up for each new project. This can use material from any source – the master component library, from external sources, plus component information compiled specially for the project in hand. The accumulation of all this data is not a task that necessarily has to be completed before any model assembly can begin. Nevertheless a start on it should be made by copying useful components from elsewhere and by creating new ones, as necessary.

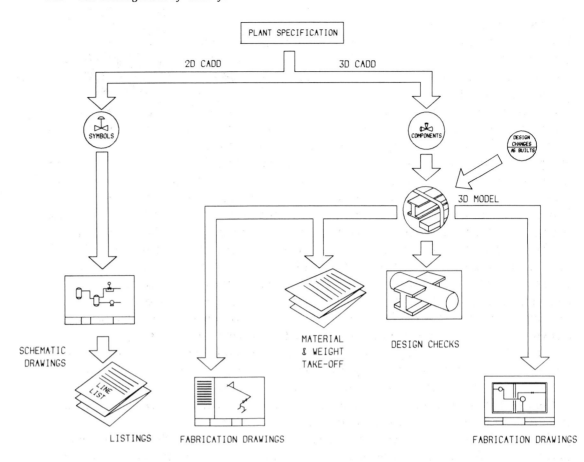

Fig 10.2 A blend of 2–D and 3–D for documentation of plant design. (*Courtesy*: Brown & Root, Vickers. London.)

A grid system

The purpose of a grid system is firstly to provide a geometrical framework on to which components like columns and wall partitions can be positioned accurately during the building assembly operation. It is also a spatial reference system used when manipulating design information and when requesting drawings.

Since the building models with their components are assembled using full-size dimensions to 1 mm precision, the grid itself must be set up at full-scale. A *planning grid* is useful for this purpose and this would normally be an evenly-spaced square grid, perhaps with a 300 mm spacing.

Superimposed on this, it can be beneficial if a *structural grid* can be specified. Many buildings require an irregular structural grid however. So this might be orthogonal with irregular grid spacings, or angular.

A *survey grid* with wider spacing can be superimposed too. This is needed for large projects, and particularly when several buildings are to be located on the site.

A 3–D locational model or any other form of 3–D model might require in addition a series of levels to represent the third dimension of its grid system. These provide the vertical control for placing components like window lintels at their correct level. They identify the main horizontal planes within the design model. Either the structural or finished levels of floors are normally used, although foundation levels or levels of suspended ceilings will also be adopted as appropriate.

Component placement

Now we can start to assemble the model. This is done by calling up components one at a time from the project component library, and then placing instances of them into the required positions in the project. Any component instance may be subsequently repositioned, deleted or replaced.

One procedure is first of all to display the plan view for the appropriate level and sector of the building. The grid can be either displayed as a background, or else it may be turned off. When any component is selected, its plan view as defined in the project component library will appear on the display screen. Then the shape or symbol may be moved to any plan position on the screen, and placed accurately to a precision of 1 mm full-size. In the process, the component could be rotated through any angle and handed if necessary. Sometimes it is easiest to attach the component first to a nearby grid position and then shift it any desired number of millimetres along the two grid axes.

The component origin point is a point within the component plan view which is defined when the component is created. This same point would be shown too in any defined images representing the elevation or section views of the component in the locational model. Likewise the origin point would be defined in 3–D representations for the component where these have been defined. The origin point is therefore a vital element in every component because it facilitates its accurate positioning.

In a 3–D locational model, the component is first placed at the correct plan position on a selected level. Then it is moved by any distance upwards or downwards from this level as required. This places the component accurately in 3–D space.

This placement is a very satisfying procedure – provided of course that the CAD system in use permits the operator to drag the components around to their required position on the display screen. It is the electronic equivalent of what we do when we move little cardboard models of desks, chairs and bookcases around on a board when trying to plan a new office arrangement. The system may allow many copies, or instances, of a component like a structural column to be quickly placed at positions specified either along a line or across a matrix of points.

Any components already placed can be repositioned or exchanged for another. Any component can be deleted. Physical clashes between components are likely to show up immediately on the screen.

A variable component may be called up. With this, all the variable dimensions or variable attributes which it contains must be fixed during the placement process. Each variable dimension might for example be defined using two 'hits' with the mouse of the data tablet or digitiser. The computer then calculates the distance between the points and replaces the variable with this value. After all the variables have been defined, we have produced a new instance of the variable component. This then can be manipulated in any way just like any other component. Thus a partition may be placed to fill the gap between two columns. A staircase may be placed to run between two floor levels.

Component menus

A component can always be called up by keying in its unique name. However, it is often more convenient if a series of related or commonly-occurring components can be combined together within a menu. Once such a menu has been defined, this can be displayed on the screen. Alternatively, a menu may be plotted on paper and then this can be taped to the surface of the digitiser. Either way, any one component may afterwards be selected from the menu simply by pointing to it with the mouse or digitiser stylus.

CAD systems which have this facility usually allow a component menu to be defined with a menu name, a description and a finite number of pre-defined icons or symbols to represent components. At any subsequent time the menu itself may be amended, added to, displayed, listed, printed, saved, or copied to or from other projects. In practice, a series of menus can be defined in this way. Then the operator can call up any one menu by name and select any component appearing on it.

The CAD co-ordinator or senior operator must see that components made available on such menus are arranged in some logical fashion. The retrieval of components from menus should operate in a manner which is consistent and predictable. This will help to make the whole system easier to use, particularly for infrequent users.

Viewing the model during assembly

During assembly we continue to work with full-size dimensions. Indeed scale hardly needs to be considered until later when drawing sheets have to be defined.

When working at the display screen, we can move out so that a wider area is displayed at small scale, or zoom in close so that a small area of the project fills the screen. The precise scale at which the project is displayed on the screen is immaterial. At any time the operator will want to pan across or switch to display another area of the project. All such panning, zooming or switching operations are used frequently and must work fast. The computer must not lag far behind the operator's train of thought. Speed of operation of course depends on the nature of the CAD software, the power of the hardware and the volume of the current workload in a multi-tasking computer. Computers which have to operate simultaneously as host to several users sometimes become too overloaded. Depending on the nature of the work, personal computers sometimes are not powerful enough even for one user.

At one time it was thought that a relatively slow response would be acceptable because operators needed thinking time. However research[33] has showed that rapid system response, ultimately reaching subsecond values, offers the promise of substantial improvements in user productivity. Indeed productivity increase was found to be in more than direct proportion to a decrease in response time.

In some systems and with adequate 3–D information assembled in the model, the user can switch freely from plan views to sections views, elevations or perspectives. Indeed it may be possible to carry out assembly with several views of the building displayed simultaneously on different parts of the screen.

Each of such displayed views covers a defined part of the model. It is sometimes known as a *window* into the design model. The rectangular area of the screen on which a window is displayed is called a *viewport*.

The simplest case of all is of course when one window is defined and the one viewport includes the whole screen. However in 2–D modelling, one window might be defined to contain an overall view of the whole extent of the project, while a second contains a local area of the model where detail design work is progressing. Both views can be displayed simultaneously, and are updated continually by the computer. It follows that these views must be consistent with one another. If a component is moved, it is moved in the model, and all views of the model will automatically indicate the change.

By maintaining two viewports, the designer can to keep the overall context of his work in view, while simultaneously being able to get close enough for detailed design work. In 3–D modelling, two windows might contain overall and close-up plan views, while a third window might contain either a sectional-elevation or perhaps a perspective view. Clashes between items being placed should show up clearly.

Obviously multi-window facilities provide the operator with an enhanced degree of control over his work during the assembly process. Good screen resolution is important but unfortunately the physical size and resolution of all graphics screens is limited.

The locational model

This model contains information specifying the location of every instance of the assembled components. The actual definition of each component remains stored within the component library or catalogue.

This type of model has the advantage that the model itself is compact and in turn this means that the computer processing and storage can be handled efficiently. For example, a component like a particular door-set may occur many times throughout various levels of each of several buildings within a project. No matter how many instances there are in the assembled model, the graphics or geometry of this door is defined once and one copy only is needed for the whole project. Each instance basically requires only a record to be kept by the computer of the component's name, and its position and attitude in 3–D space.

When a plan or elevation of the building is to be displayed, the appropriate image in the component library is accessed. The computer displays this same image at all the positions where instances occur within a view.

Component modification or exchange

When a component is stored once and several instances occur in the design model, there are various implications of interest:

(1) If the component definition is modified within the component library, all instances of the component within the project will be automatically updated with the one modification.

(2) If a new component is exchanged for the first in the component catalogue, the new one will automatically replace the first on all views and drawings subsequently produced.

(3) When the name of one component instance is changed to that of another available component, only that instance is swopped.

When components are exchanged for items which are physically larger or smaller, the operator must be aware that the adjacent components are not adjusted. He must organise all the modifications which are made necessary as a consequence of the first change, and watch out for physical clashes.

10.4 DESIGN BY REPLICATING AND ADAPTING UNITS OF DESIGN

So far in this chapter, I have concentrated on the building up or assembly of a design model from constituent CAD components. This 'bottom-up' approach is commonly adopted, but is certainly not the only way in which design solutions can be produced.

Top-down design

A previous design solution might be retrieved and used as the basis of the new solution. Either the previous work could be used essentially as it now stands, or else it could be taken and adapted to a greater or lesser extent to suit the new circumstances. This adaptation of an existing solution, we might term as the 'top-down' approach to design.

Clearly there are many types of buildings which recur in different locations. They are rarely identical if only because the foundations and landscaping must be different. In these circumstances it would be easier to replicate the previous design solution, and then work to adapt it as necessary for the new location. Examples occur in the design of multiple retail chains, supermarkets, banking, and residential accommodation.

In CAD terms, this is achieved by first making a copy of the previous design model, and transferring the copy into the space allocated within the computer for the new project. This can always be done quickly and precisely by the computer. Then the full manipulative capacity of the CAD system is brought to bear on this copy to adapt it until it suits the new requirements. When there are opportunities, this 'top-down' approach is a very powerful means of design. The CAD co-ordinator, liaising with project managers, must always be on the look-out for the opportunities.

Figs. 10.3 to 10.6 and Plate 5 illustrate the top-down approach. Fig. 10.3 is a plot of the survey for the site where a new petrol filling station is to be constructed. Fig. 10.4 is the plan of a standard layout for a filling station. This represents the standard design that is stored permanently within the computer. Fig. 10.5 is this standard design overlaid on the site survey. Using the editing facilities of the CAD system, the designer can soon modify and tidy this up. The object is to produce Fig. 10.6, the design for the filling station that is to be built on this particular site. Where the main components are defined in 3–D as well, it is possible for the computer to provide 3–D views of the new design (Plate 5).

'Sideways' design approach

Another design approach is to identify some part of the current assembled project and then make one or several copies of this unit for transplantation elsewhere. If this is all done within the same design solution, we might refer to this as the 'sideways' approach to design. The unit which we would pick up, copy and transfer could be a room or group of rooms in a building, a whole floor level in a multi-storey building, or perhaps even an entire building.

In the CAD transfer process, the design unit could be rotated or handed about any convenient anchor point, and then shifted. The destination could be literally any appropriate position within the same project.

This too is a very powerful facility. It is when this replication process is combined with the always-available means of manipulating and adapting the copied unit, that the full power of the approach can be perceived.

Many projects have design units of some kind that are repeated several times, albeit with some adaptations. Much can be achieved if units of layout that already exist are taken and modified to fit a new purpose.

Some examples of projects where this replicate and adapt facility ('sideways' design approach) could be profitably used within the one project are:

Project type	Typical unit that might be designed once & then be replicated & adapted
Hospital	Ward
Hotel	Bedroom
Office building	Work unit
Apartment block	One apartment of each type
Housing estate	One house
Multi-storey building	One typical floor
Factory	One production unit

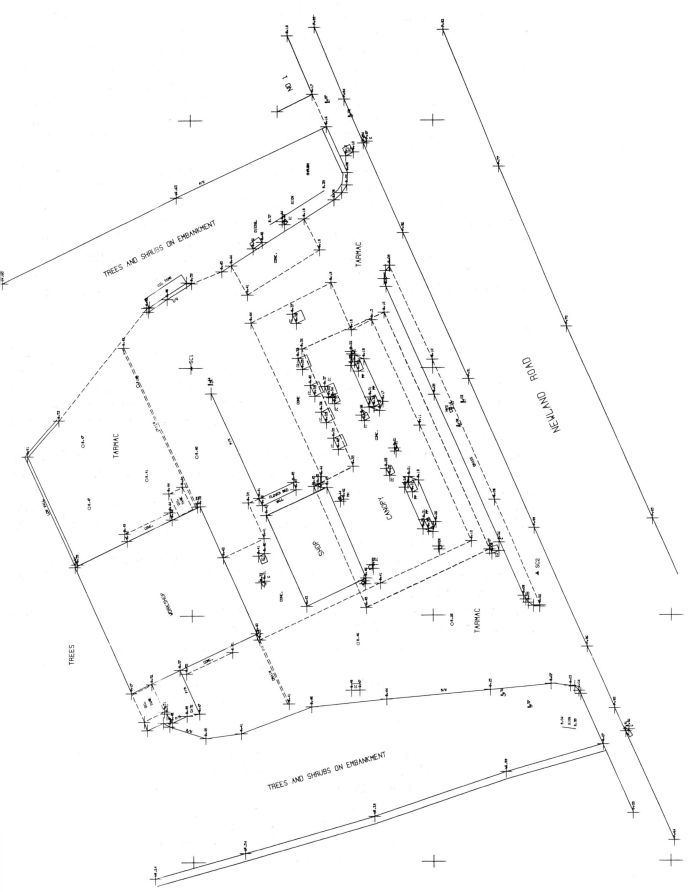

Fig 10.3 Site survey plan for petrol filling station. (*Courtesy:* Shell UK Oil Retail Division. London.)

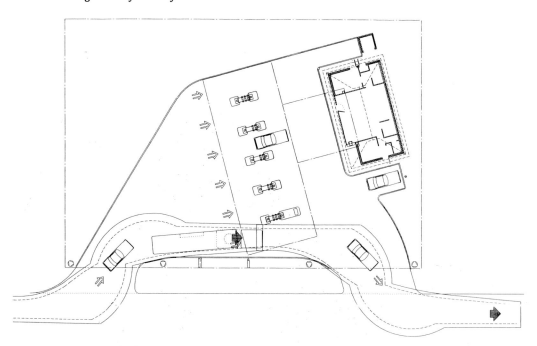

Fig 10.4 Standard retail layout plan. (*Courtesy*: Shell UK Oil Retail Division. London.)

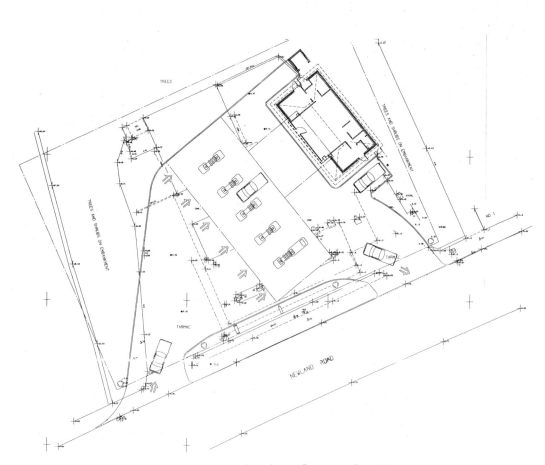

Fig 10.5 Standard layout overlaid on site plan. (*Courtesy*: Shell UK Oil Retail Division. London.)

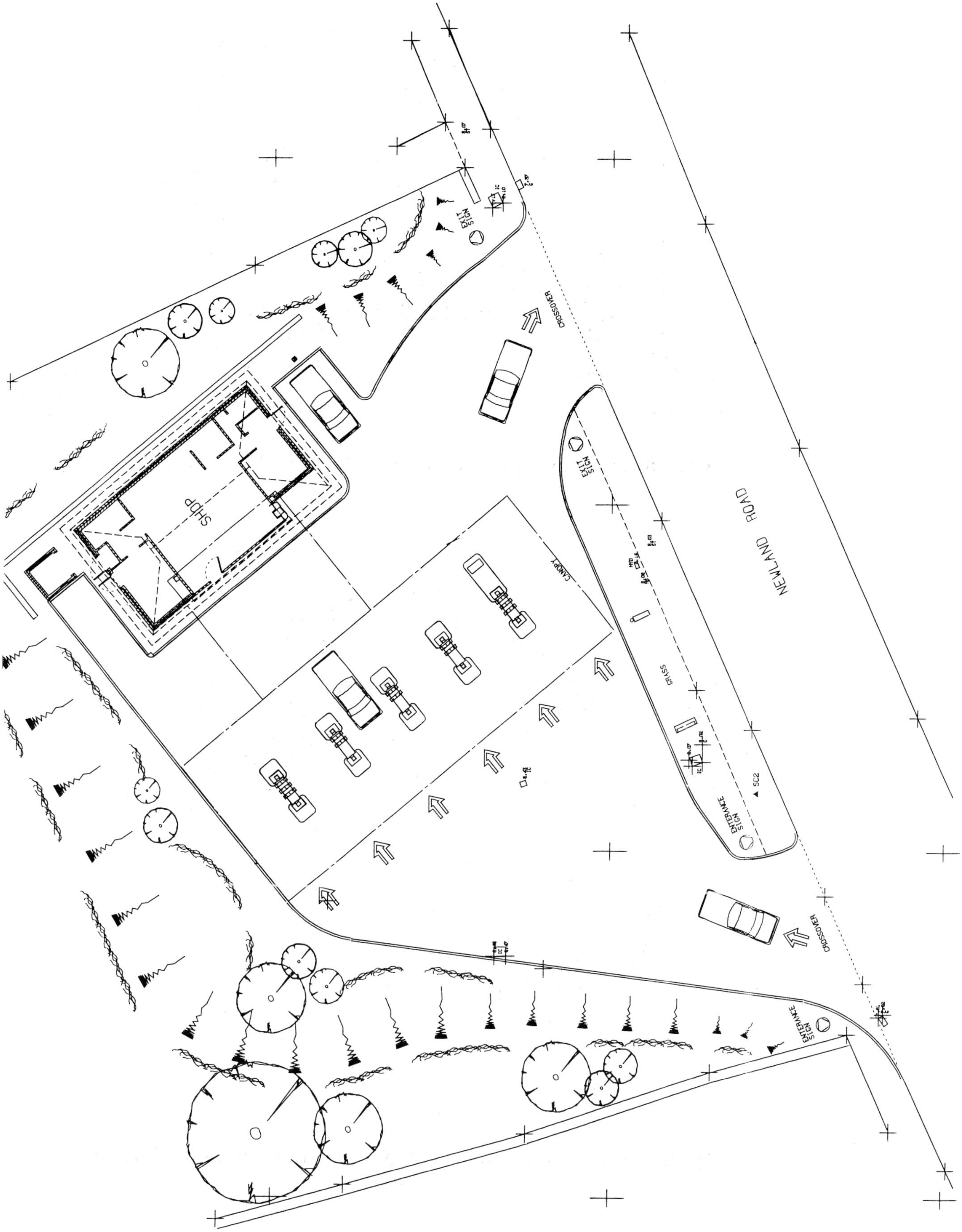

Fig 10.6 Standard layout modified to suit the site. (*Courtesy:* Shell UK Oil Retail Division. London.)

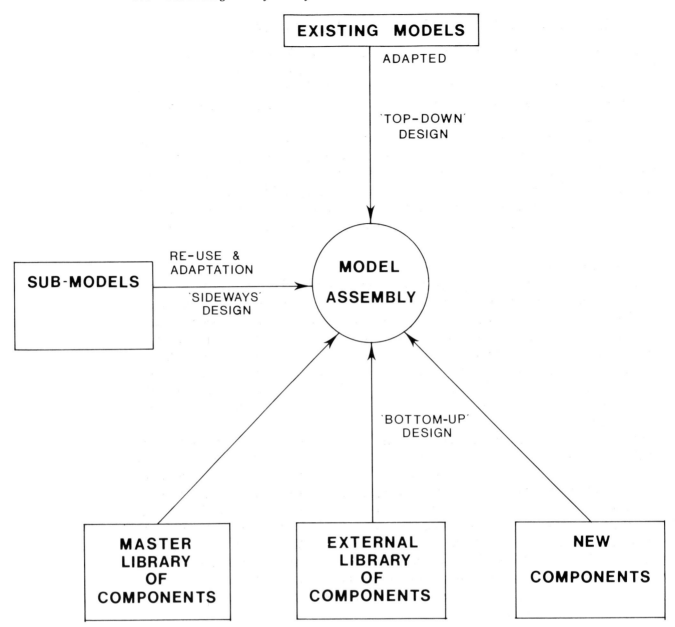

Fig 10.7 Diagram of model assembly process.

'Bottom-up', 'top-down' and 'sideways' techniques are all different approaches to design (see Fig. 10.7). There can be opportunities for all three and certainly CAD modelling can benefit if a mixture of all these approaches is adopted. Potentially reusable components, reusable design models, and reusable units lying within the current design model – are all valuable assets within the design office.

It is far better when these are saved within a computer store, rather than merely on paper. The computer's ability to make identical copies, and the CAD system's ability to move and adapt them is immensely valuable. Indeed this is an indication of how a design organisation is able to build up its expertise in a tangible, machine-readable form. Over a period, such reuseable design material can be built up both in terms of quantity and quality. With their reuse it becomes possible to turn out new design solutions with considerable speed and with greater confidence regarding quality.

1O.5 CLASSIFICATION OF DESIGN INFORMATION

We have seen how with a component-based CAD system a design model can be formed from the assembly together of a very wide range of components. Even when we transfer large units of design information from other projects or from other locations elsewhere within the current project, the design model still remains basically an assembly of CAD components.

Let us now look at what happens when we come to extract information for the drawings and schedules. It would certainly not be convenient if we had always to plot literally all the available design information in the model, or even that within a specified zone of the model. Far too much information would appear and it would be of little use to anyone. It is essential to have control over the nature of the design information which is plotted. For this we depend on the data classification features of the CAD system.

Component classes

In a component-based modelling system, each component could be classified at the time of its creation. This class or component category then remains an inherent property or attribute of the component, and likewise of every instance of the component which is assembled into a design model. For example all doors might be assigned to one class, windows to another, internal partitions to a third, external walls to a fourth, ventilation ducts to a fifth, furniture to a sixth, and so on. Allocation of component classes is done by the system user and so is fully under his control. Such classification can be as detailed as required. Thus we could assign concrete columns, steel columns and brick columns to three distinct classes.

Once allocated, this form of component class is envisaged as a semi-permanent attribute of the component. Nevertheless to provide flexibility and to cater for special cases and for errors, it is essential to be able to change the class of any component at any time.

Operators must have a classification coding scheme set up for them by the CAD co-ordinator so that they know how to code each new component. This must be done immediately the new CAD system is implemented. Ideally the scheme would be suitable for all projects. However we should expect that the particularities of some projects may well mean that some additional project-specific classes might be needed from time to time.

A small part of such a component classification system could be:

Classification code	Nature of component
430	External doors
434	Internal doors: 0.5 hr fire rating
435	Internal doors: 1 hr fire rating
436	Internal doors: 2 hr fire rating
620	Light fittings
621	Electrical fittings: power control
622	Electrical fittings: power distribution

When a drawing is to be extracted from a design model, we could define any combination of component classes to be included or omitted. This means that the drawing could depict any set of construction elements. For example a plumbing drawing could depict the plumbing fittings, sanitary appliances, partitions and concrete slab floors, together with outlines of the holes in these slabs.

Another application of component classes is to allow switching between different graphical representations of the same component. For example a CAD component for a door-set might be created with a very simple image and defined under one component class. Another CAD component with another name could be created to represent the same door-set. This second one would be provided with much more detail in its graphical image and allocated to another class. Two CAD components now exist for the same real component, and both CAD components would be placed at identical positions in the model to represent the one door-set. This is a device, of course. Its value arises because we could define a small-scale drawing and request the first component class to appear, but not the second. At another time we could define a large-scale drawing and request only the second component class – the one which shows the greater amount of detail.

Design layers

Many drafting systems adopt a system of drawing layers for classification purposes. Anything drawn in is stored on the layer with is currently set. So structural columns might be drawn first and saved on one layer; then another layer is switched on and the partitions are drawn. Each layer with its drawn information is analogous to a single transparent plastic overlay.

When a drawing is required, it is possible to specify that any combination of layers should be switched on. What is displayed or plotted is what would be seen if only the equivalent overlay sheets were combined and viewed. By defining a series of different combinations of layers, it is possible to extract a series of drawings from the same source of drawing information. Each of the drawings would individually have contents to suit specific purposes.

In a component-based modelling system the *design layer* is a means of classifying each *instance* of a component. The design layer is distinct from the inherent component class or category. The design layer is assigned not when a component is created, but when each instance of a component is assembled into the design model. So the design layer classification system can be specific to a project, and can reflect the characteristics of it. It enables us to record the *use* or *function* of elements within a particular design.

Here are some more examples of the use of design layers:

- *Phasing*
  For distinguishing different construction phases
  Existing site features and new construction
  Features to be retained and removed
  Temporary works and permanent works
  Refurbishment – before and after
- *Zones*
  For public areas
  Airside and groundside in an airport
  Offices and retail areas etc
- *Ownership/departments*
  Furniture and fittings for administration, management, drawing office, etc.
  Adjacent properties
- *Function*
  Load-bearing structure, non-load-bearing structure
  For classifying services

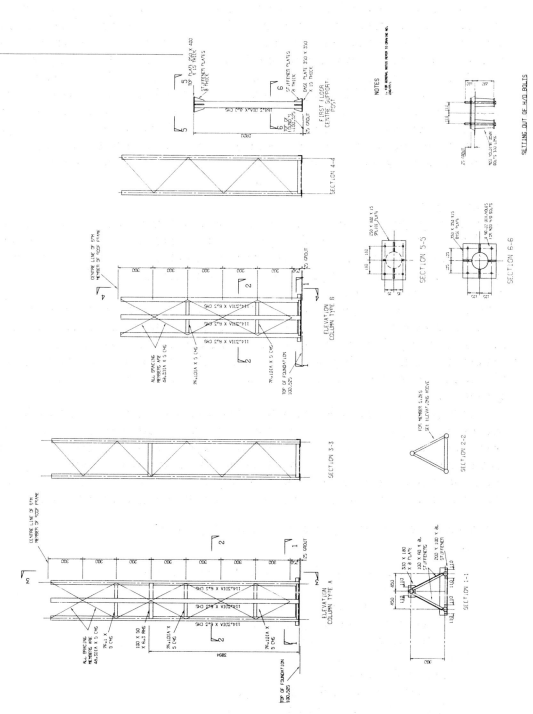

Fig 10.8 Steelwork details New Passenger Ferry Terminal Sheerness. (*Courtesy:* Blue Chip Computer Services, Dartford, England.)

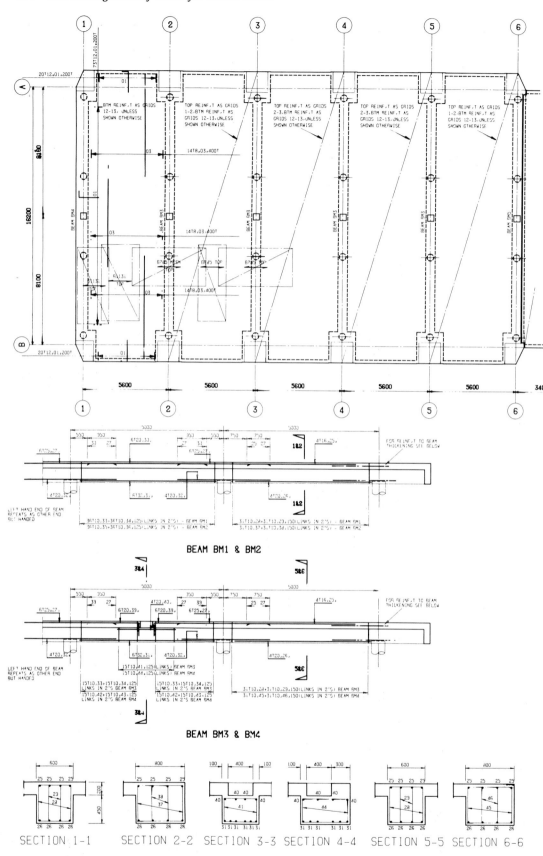

Fig 10.9 Reinforced concrete details New Passenger Ferry Terminal Sheerness. (*Courtesy:* John Allen Associates, Dartford, England.)

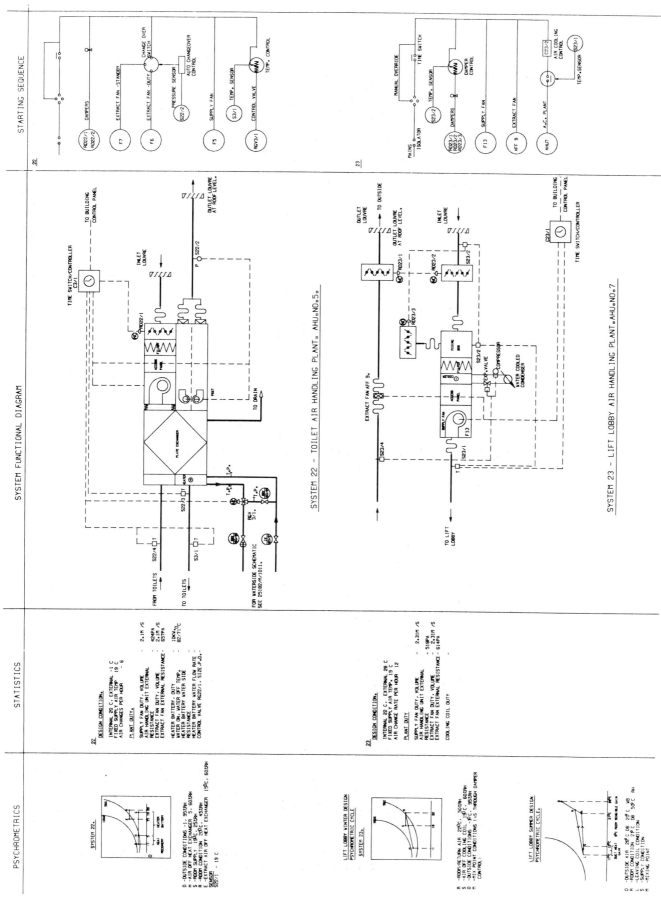

Fig 10.10 Air plant functional diagrams. (*Courtesy:* Haden Young Ltd. A BICC Company. Southgate, England.)

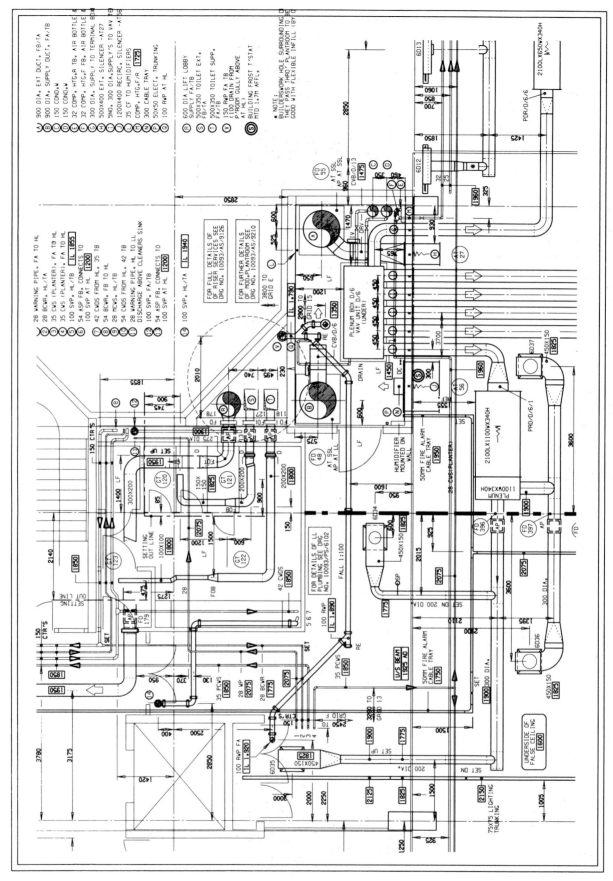

Fig 10.11 Plant room layout. (*Courtesy*: Haden Young Ltd. A BICC Company. Southgate, England.)

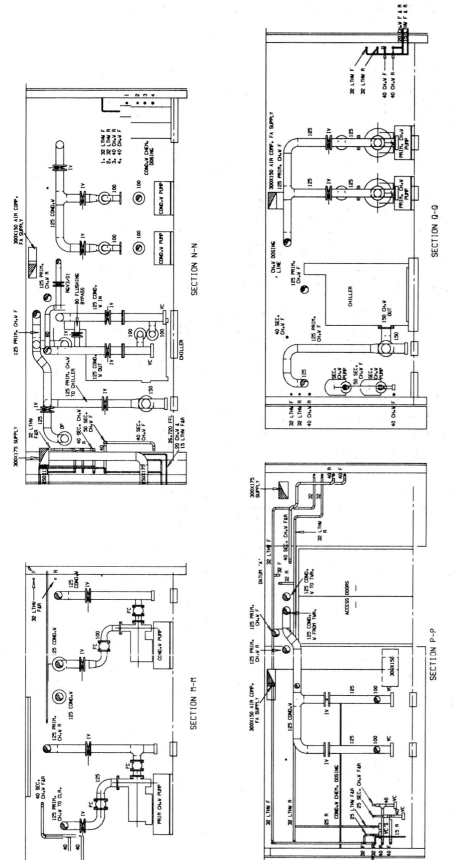

Fig 10.12 Plant room sections. (*Courtesy:* Haden Young Ltd. A BICC Company. Southgate, England.)

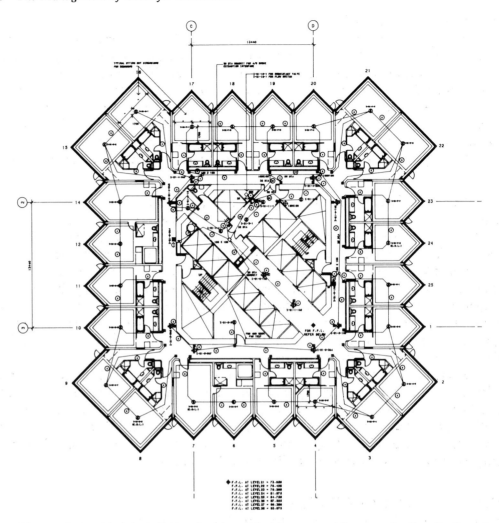

Fig 10.13 Fire services installation in a hotel development. (*Courtesy*: Jardine Engineering, Hong Kong.)

- *Alternatives*
    To distinguish between alternative design proposals
- *Revisions*
    Initial design
    Revisions A
    Revisions B
- *Specification*
    Engineering brickwork, facing bricks, blockwork
    Fairfaced concrete, exposed aggregate concrete
    Gloss paint, emulsion, etc.

The content of drawing sheets can be controlled by these means. Figs. 10.8 to 10.13 illustrate drawing sheets containing steel, reinforced concrete and building services details.

Dating of components during assembly

Another means of classification is built into some CAD systems, and it operates automatically. At the end of each assembly work-session on the system, every component placed during the session is tagged with the current date. If an existing component was moved during the session, it is redated. Components which are deleted – are strictly not deleted from the design model at all – but are dated and marked as not-current.

The purpose of this dating function is to enable a drawing to be requested which shows any additions, alterations, or deletions to the design model between any two selected dates. It also provides a means of automatically recording progress of the design work.

Classification of design data and project management

Component classes, design layers and component dating can be used in conjunction. Together they enable the design model to be structured by the user. Used in combination they provide powerful facilitates with which a design project can be managed.

For example, a set of components could be assembled to form a run of ductwork and assigned to design layer 1101 to represent the function 'air supply'. Identical components – and therefore having the same component class – could be assembled into another run of ductwork. However this run might be assigned to design layer 1102 to represent 'air return'. In this way, we have the ability to classify services, or indeed any other design data, in a variety of ways.

I have explained these classification features at length because of the strong bearing which they have on project management and on drawing definition. Clearly the CAD co-ordinator and users have to give thought to their application during component creation and building assembly phases. For example in Chapter 12 we will discuss the importance of controlling design revisions. We can see now that design layers and component dating provides some facilities for this control.

10.6 EXTRACTION OF DRAWINGS

In this section, I shall indicate how drawing sheets are defined by a CAD system. The procedure is usually closely related to the classification of the data in the model.

First, the distinction between a 'drawing' and a 'sheet' should be understood.

A *drawing* is a view of a part of the computer model. For example, one drawing might portray the plan of a floor of a building at 1:50 scale, with plumbing details included. Another drawing may be a 1:20 vertical section view cut through part of the same floor, again showing the plumbing details. A third drawing might be a small-scale perspective of the whole building, used as a key drawing. Each drawing is a defined 'window' looking into the design model.

A drawing *sheet* is the paper equivalent of the display screen. It is an area on which one or several drawings are included. Thus an A1 sheet might be composed with the three drawings already mentioned set out inside an outline frame and with a suitably labelled title block. In practice, the 'sheet' is composed within the computer and can be viewed at reduced size on the graphics screen. When plotted, it becomes a real sheet of paper or plastic film.

Drawing definition

A plan drawing would be defined in a modelling system by inputting a drawing reference (or name) and the required scale. Since all model data is held in full-size dimensions, it is easy for the computer to rescale to any desired value. The limits of the area of interest are defined perhaps by indicating with the mouse the positions of the bottom-left and top-right points (as seen on the display screen). Next the component categories and design layers to be included must be defined. Further information is needed regarding the background grid to be drawn, if any, and the text which is to appear as annotation. All this information together constitutes a '*drawing definition*'.

In Chapters 3 and 9, I explained that a CAD system holds some kind of graphical representation of each component. This might be a single 2–D image, several 2–D images or some form of 3–D representation based on lines, surfaces or solid elements. Using the parameters of a drawing definition, a CAD system which uses a locational model can determine which of the many components will come within the limits of the drawing window. The appropriate representation of each component is selected. If the drawing is a plan, elevation or section, then the relevant diagram or symbol is retrieved. If the drawing is a 3–D view, then the image of the component is generated from the 3–D geometry. The drawing is the sum of all relevant component representations, each plotted at locations defined in the locational model of the building.

Each time the drawing is requested, it is generated using the information in the current version of the design model.

The parameters which define each drawing can be saved as a unit. This is the *drawing definition*. Each one is given a unique name when it is saved in the computer. This means that at any time in the future, perhaps after the design model has been modified, it is a simple matter to call up the same drawing definition to generate a revised drawing.

The generation of elevation and section views using a Sonata or Rucaps type of locational model was explained in Section 3.6. A drawing definition then includes the cut or section line and parameters to define a volume of interest. All components of the defined category and design layer falling within the volume will be selected. In principle this is much the same as with plan drawings, but this time it is the component elevation or section graphic representations which are retrieved and used.

3–D drawings are defined in a rather more elaborate way which includes the type of graphical projection, with parameters of the view-plane and view-point.

Once generated by the computer, such drawings can be viewed on the display screen or are available for composition within a sheet.

Sheet definition

Any number of drawings, each specified by its drawing definition name, can be placed in position within the sheet. Positioning could be effected with the aid of the data tablet or mouse. In addition to these drawings, it would be possible to include graphical components in the sheet. Examples of these would be a sheet frame, a title block, revision box, blocks of text, text for drawing titling, north points and so on. All the parameters which together define the sheet can be retained within the computer as a unit too. This is the *sheet definition*, and this would be given a reference name and a sheet size, such as A1.

Figs. 10.14 to 10.17 illustrate the composition of sheets, which can include plans drawings, elevations, sections and 3–D views. It is the 3–D views that are unusual in traditional manual drawing practice. They enhance the clarity of the overall sheet and improve communications.

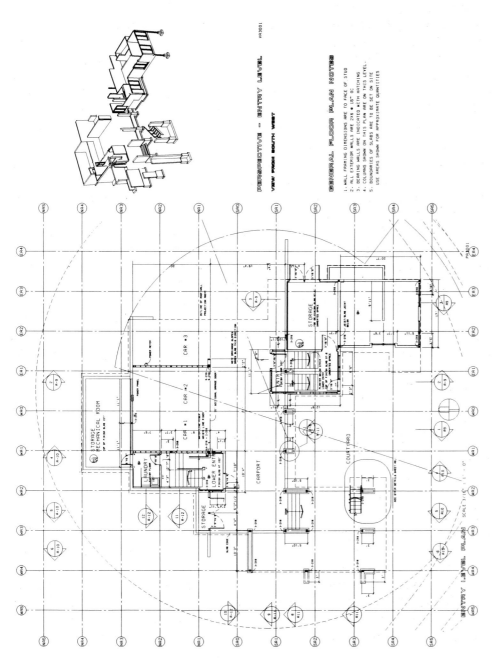

Fig 10.14 House general floor plan. (*Courtesy:* GMWC North America. Robbie Wakeling.)

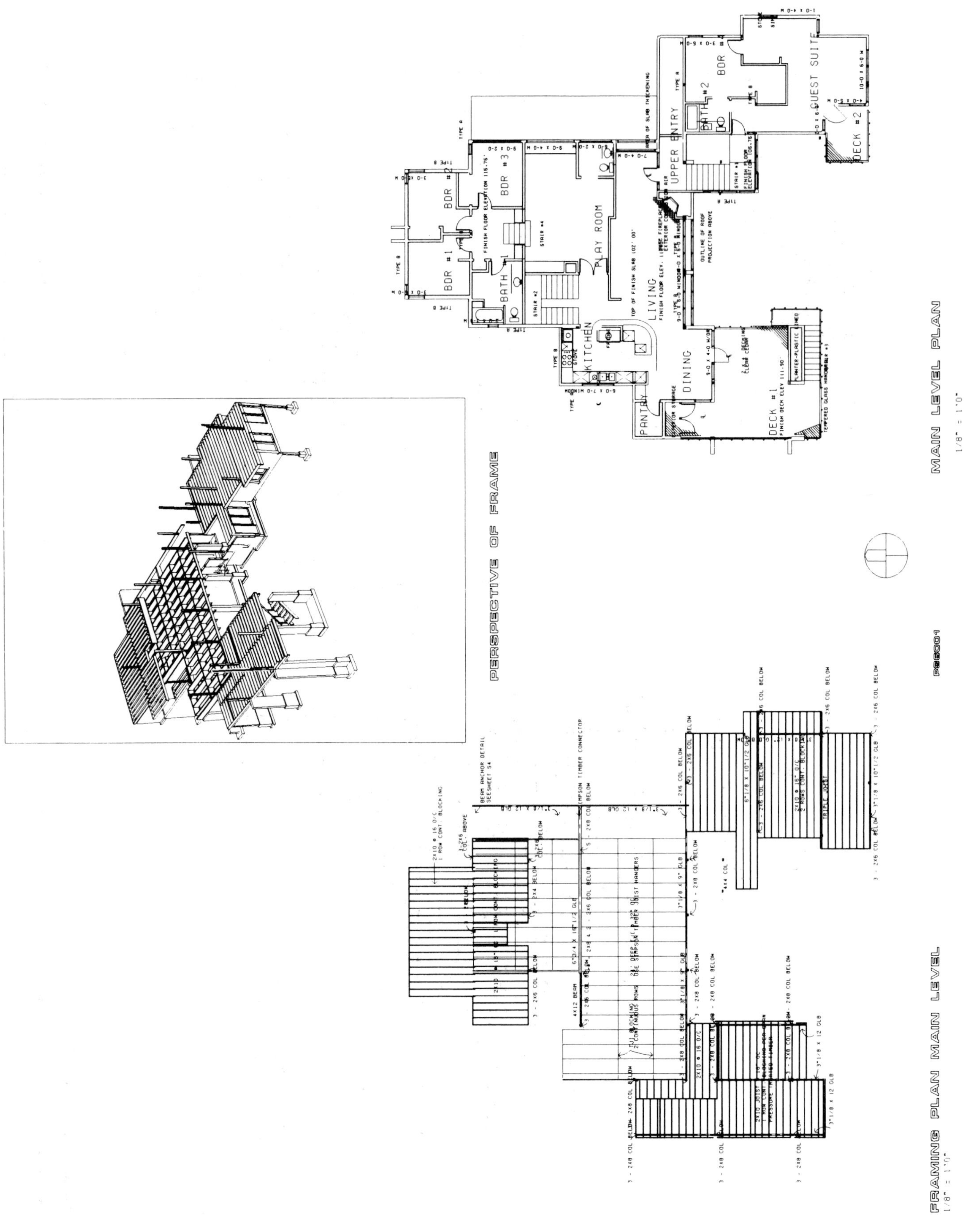

Fig 10.15 House framing plan. (*Courtesy*: GMWC North America. Robbie Wakeling.)

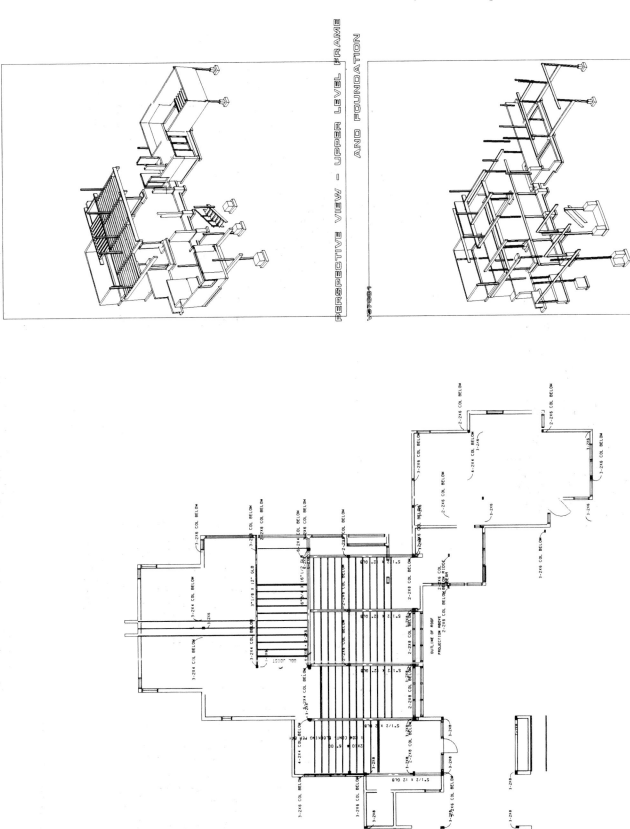

Fig 10.16 House framing plan. (*Courtesy:* GMWC North America. Robbie Wakeling.)

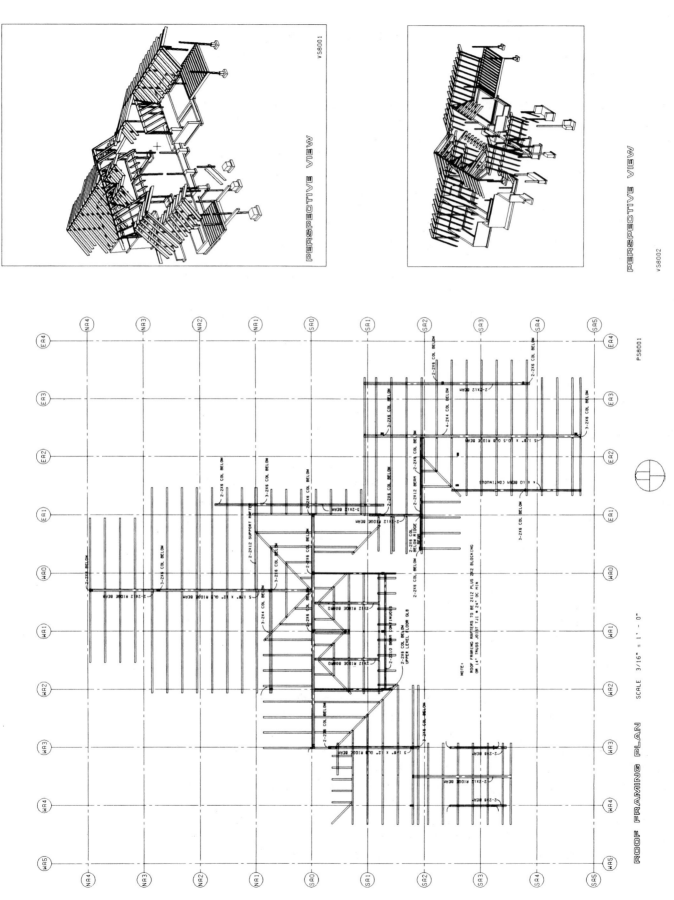

Fig 10.17 House roof framing. (*Courtesy:* GMWC North America. Robbie Wakeling.)

Two methods of sheet extraction

The general principles of drawing and sheet definition have been explained. It is important to note, however, that different CAD systems adopt differing courses in the extraction of data for sheets:

1 Sheets are extracted from the model and no subsequent embellishment is possible.

   Here the model has access to all the design data to permit the sheet to be composed. Component images are stored or can be generated automatically. All annotation and dimensioning is included within the model. Drawing definitions and sheet definitions are likewise stored. When the plotting of a sheet is initiated, a finished sheet is automatically generated on the plotter.

   Several sheet definitions will be saved in the computer. Then after the design model has been modified, any revised sheet can be plotted simply by calling up the relevant sheet definition. The computer handles the regeneration of each of the drawings and the composition of these into the new sheet. Each sheet is like a snapshot of the model and it accurately reflects the state of the model at the time of its production.

2 Drawings are extracted from the model. Then the user composes and embellishes these to form a sheet.

   Several views of the model (i.e. drawings) are extracted from the model. These can be composed within a sheet frame on the graphics screen. There is no sheet definition as such. At the outset, the drawings reflect the content of the computer model. But additional graphics, dimensioning, annotation, titles and so on (which are not within the model) can be added using editing methods provided in the system. The snapshots of the model can be embellished before use.

Method 1 has the important advantage that revised sheets can be produced quickly and easily after any revision has been made to the model. The sheets always accurately reflect the contents of the model, and multiple sheets will be consistent with each other. It has the disadvantage that everything must be carefully input and stored within the model or in the sheet definition. It is sometimes not easy to deal in this way with annotation, dimensioning, or minor variations within the design.

Method 2 had the advantage that sheets are easier to compose. A graphic detail can be added here, and extra few lines added there, an extra few notes or dimensions are easily inserted. The problem is that when the model is revised, the embellishment might not fit, or be appropriate to the new situation. After revision then, the embellishement may have to be revised too, and so it is a less automatic process.

Traps for the unwary

There is a trap to be avoided at all costs – and this applies with both methods and to any modelling system. No plotted sheet should be altered manually – at least not without making the corresponding change in the computer. In pressure situations manual changes sometimes are made and the model is not updated. If this is done, then the plotted sheet no longer reflects the state of the model, and this leads to considerable confusion.

Another trap for the unwary is to issue or use sheets which individually have been extracted at different times for a design model which in the meantime has been revised. The sheets will not be consistent but this may not be obvious. One guard against this happening is to adopt a system of updating the name or reference of models and sheets after design revisions have been made. (see Section 10.7 and Chapter 12.)

10.7 MODEL AND SHEET REFERENCES

Every design office has a scheme for naming its drawing sheets. This operates for manually-produced drawings and perhaps is well ingrained.

When CAD is introduced, it is certainly important that each computer model and CAD-produced sheet should be assigned a unique reference or name. The scheme adopted for sheet names should be applicable for manual and CAD sheets. There is no universally-acceptable practice in this matter, and indeed every office tends to have its own preferences. It is recommended therefore that any new scheme for CAD should reflect and ideally blend with the existing conventions – there is rarely any need for a revolution! The following ideas for model and sheet referencing are put forward merely as examples, or for guidance where existing practices are not working well. The CAD system in use may impose limitations.

Model name

The model reference should be unique, and should reflect the design revision and current status of the model. The name code could be compiled having the following fields (or a subset of these):

- Project number   e.g. 4 digits
- Model number   e.g. 2 digits
- Control status   see Chapter 12 for an explanation.
  e.g. 1 character, viz:
  - W = Work-in-progess (to design)
  - C = Awaiting checking
  - A = Awaiting approval
  - R = Released for use
- Model Revision Code   e.g. 1 character
  to reflect model revisions after having been released.

Example: 1234/03RA

This is revision 'A' of model '03' of project 1234. It has 'Released' status.

Sheet names

The sheet reference should be unique, and should reflect the design revision and current status of the model and sheet. The name code could be compiled having the following fields (or a subset of these):

- Project number   e.g. 4 digits
- Model number   e.g. 2 digits
- Control status   e.g. 1 character, viz:
  - W = Work-in-progess (to design)
  - C = Awaiting checking
  - A = Awaiting approval
  - R = Released for use

- Model Revision Code   e.g. 1 character
                               to reflect model revisions after
                               having been released
- Design discipline of sheet   e.g. 1 character, viz:
                               A = Architecture
                               S = Structural
                               M = Mechanical Services
- Sheet number   e.g. 2 digits

Example: Model No. 1234/02CB/E012

This is revision 'B' of electrical services sheet number '012', extracted from Model '02' of project 1234. It is sheet plotted for checking purposes.

CHAPTER 11
Non-graphical Design Information

In previous chapters we have studied how graphical and geometrical design information can be represented in CAD systems. Drawings can either be produced by computer drafting techniques or extracted from design models.

Design information is, of course, not limited to graphics and so we need a means of handling non-graphic information too. However computers have to cope with the two sorts of information in different ways, although ideally a link is required to associate the non-graphic attributes with their related graphics.

This chapter introduces the subject of database management systems. The relational type of database is suitable for handling non-graphical attributes and this is discussed. Typical attributes of CAD components, of design units, drawings and sheets are all described.

The process of building up a non-graphical database for a project, and of linking this to the geometrical layout model is then tackled. Then the various kinds of reports, schedules and costings can be produced.

Some CAD systems have no facility for linking graphical and non-graphical data. Then of course the two kinds have to be handled entirely separately.

11.1 NON-GRAPHICAL DESIGN INFORMATION

We have now seen how components can be defined in terms of their graphical representations. After a component instance is assembled into a design, it will show up on displays and drawings either as a symbol, a 2–D outline or diagram, or as a 3–D view such as an orthographic projection or a perspective.

Text information can be included as a *part* of the component graphics. Then it too will appear as annotation on the displays or plotted drawings where the component has been placed. A block of text information can even be treated as a CAD component and included within the geometrical model. Alternatively some CAD systems allow a drawing sheet to be extracted from a geometrical model and then this can be embellished with additional annotation. Nevertheless all these methods cope only in a limited fashion with non-graphical design information.

We must be clear that design information is a complex mixture of graphical and non-graphical information. Unfortunately computers cannot handle large volumes of graphical and non-graphical data by using the same methods. So what has happened is that CAD systems have been developed which concentrate on dealing with the graphical/ geometrical information. They place this data in graphical databases and the software is carefully 'tuned' for manipulating and displaying this kind of information. This must be so if the system is to operate efficiently.

There is no point in placing large quantities of non-graphical information in the graphical database, because the system would not cope. Therefore, in addition to the graphics facilities of a CAD system, we ideally require the following:

(1) A means of manipulating the non-graphic attributes of:
- components,
- various units of design (e.g. parts of projects),
- whole projects,
- drawings,
- drawing sheets.

(2) A means of linking these attributes with the graphical data.

Manipulation of non-graphic attributes includes the inputting, saving, revising, copying, retrieving and reporting (e.g. in the form of schedules) of the non-graphic attributes.

The 'units of design' referred to above can be any sub-divisions of a project which are of relevance to the designer. Examples could include a room, a floor of a building or a service core. The relationship between CAD models, drawings, and parts or components was highlighted by Macdonald[34].

The graphical information is required for planning and designing the layout and assembly of a project.

The non-graphical information has more relevance for scheduling, costing, procurement and for project management purposes.

11.2 DATABASE MANAGEMENT

Database management is a subject which is much talked about, but rarely fully understood. First a few definitions are required.

A computer *file* is a set of data records in computer-readable form. These records are accessible as a unit within the computer.

A *database* is a term which is applied to an organised group of computer files. Scott[35] discusses the characteristics and difference between conventional file organisations for computers and databases. He refers to one of the key properties of databases. This he states is the integration of the data files by establishing bases of association between records both within a file and in different files.

Each proprietary CAD system builds and maintains a database for all the graphical information associated with a drawing or project design. The data structure of this graphical database is normally unique to the CAD system. It would be highly tuned so as to be especially efficient at manipulating graphical information. This structure is unlikely to be equally suitable for handling non-graphic data. It is for this reason that a modern CAD design system can benefit from having two databases:

(1) a graphical database;
which preferably should be linked to:
(2) a database for non-graphical information.

A *Database Management System* (DBMS) is a computer software system. It is a database structure plus programs to manage the database. A DBMS normally provides extensive facilities for input, storage within the database, updating, retrieval and reporting of data. Updating a record would result in the automatic updating of all related records as well.

Generally a command structure or query language is provided to permit the user to operate the system and to make enquiries. A range of reporting facilities are provided. These cater firstly for the generation of routine reports or schedules to show the data content at any time. Secondly they can generate *ad hoc* reports. This last facility allows the user to design the format of his own reports or schedules to suit any special purpose.

A DBMS will also have software to control access by users to the data and to provide database security. Application programs can be written which can obtain their data from, and submit results to, the database.

11.3 TYPES OF DATABASE

Scott discusses four types of database suitable for non-graphic information, namely:

- inverted list,
- hierarchical (tree structure),
- network,
- relational.

The *inverted list* includes a sequential file of records and in addition it maintains attribute lists. It is not well suited to applications like CAD where the data is continually being modified.

The *hierarchical* organisation can handle relationships between records where they are chained like in a family-tree. Thus it can handle the manner in which simple components are arranged into assemblies; then these into increasingly larger units of design like rooms or buildings; and ultimately these units into a complete project. It copes well with the way in which one or more items 'belong' to another.

Unfortunately the designer is not only concerned with the hierarchical arrangement of components into his complete design. Indeed he is neither concerned at one extreme with 'bottom-up' design where a project is formed by building up from components, or with 'top-down' down where a project is decomposed into components so that the design can be specified and built. The designer's approach is something more complex and lies between these extremes.

The *network* organisation can cope with more complex associations between records than a tree structure does. In theory all the associations that might exist between the numerous components used in a project could be maintained in a network. The associations are formed within the network database because the software inserts pointers within the data records. A pointer is merely the physical address within the computer of one record, placed within a second record. With such pointers, a program can proceed from the first record directly to the second, thus establishing the association between the two. Inverted list and hierarchical databases also use pointers.

Not surprisingly, there has been much interest in the network database for CAD work. Unfortunately as the number of records increases, the number of such pointers in a network database tends to increase geometrically. It can soon become almost unmanageable in practice. Another disadvantage is that the basic structure of network of associations has to be pre-established so that the pointers can be set up and maintained as records are added. We discussed in Chapter 2 that the nature of design is far too undefined and fluid to allow any rigid pattern of associations to be specially set up at the outset.

11.4 RELATIONAL DATABASE MANAGEMENT

Unlike the other database organisations, the relational database does not contain any pre-defined associations in the form of pointers which are inserted into the data records.

A *relational* file is in the form of a table. Each row of the table is a record, and the columns are fields or attributes of the records.

The associations between records are determined not by pointers or computer memory addresses, but on the basis of identical contents in the fields of different records. The related records might be all within one file, or spread across several files.

As a illustration of this, Table 11.1 contains two simple relational files. File A is an inventory of furniture in part of a hospital design. Each row is a record of the occurrence of a type of component. Each record has three fields or attributes. File B is a catalogue of furniture components, each of which has one attribute, namely the unit cost.

These two files constitute a simple relational database. The furniture costs in the hospital design can be determined automatically by relating the two files. This is done by finding identical names assigned to components. As a result, the computer is able to generate the two reports shown in Table 11.2. This is perhaps an over-simplified

Table 11.1 A relational file structure

File A: Room contents records

Room	Component	Number
Ward 1	Bed	8
Ward 1	Chair	6
Office	Desk	1
Office	Office chair	1
Ward 1	Table	1
Office	Filing cabinet	1
Ward 1	Lamp	8
Ward 1	Trolley	2
Office	Chair	2
Office	Lamp	2

File B: Component file

Component	Cost
Bed	586.45
Chair	43.54
Desk	412.22
Filing cabinet	165.66
Lamp	19.11
Magazine rack	87.67
Office chair	120.81
Table	344.69
Trolley	145.62
Washhand basin	134.56

Table 11.2 Tabular reports generated by the computer

Report 1: Total costs of furniture

Component	Number	Cost	Total cost
Bed	8	586.45	4691.60
Chair	8	43.54	348.32
Desk	1	412.22	412.22
Filing cabinet	1	165.66	165.66
Lamp	10	19.11	191.10
Office chair	1	120.81	120.81
Table	1	344.69	344.69
Trolley	2	145.62	291.24
Total			6565.64

Report 2: Room costs

Room	Room Cost
Ward 1	5741.65
Office	823.99

example but the usefulness of this technique for real hospital design can be appreciated.

The tabulated records can be in any order, and it is easy to add records or to delete others. Every processing operation results in a new table of records. It is even possible to add or change the fields within tables at any time. A table can be sorted into any order which can be dictated by the contents of any of the fields of the individual records.

The important point is that the relationships are not defined at the outset, but can evolve dynamically as the contents of the records in the database change. This provides the flexibility needed in design work.

A relational database management system is a substantial software product which requires much computer processing power and resources. A degree of data redundancy is always inherent. However this form of database is generally well suited for handling the non-graphic data associated with design work.

11.5 DESIGN ATTRIBUTES

Database managment systems are the tools for manipulating non-graphic attributes. Let us take a brief look at examples of attributes.

Attributes of CAD components

We have already seen how attributes might be attached to CAD components. In Table 11.1, File B is the table which links component names with one attribute, the unit costs. It is indeed possible to envisage certain types of attributes which apply fairly generally to a wide spectrum of components – attributes such as unit cost; overall weight; date required on site; supplier's name.

In other situations however, particular types of components require their own particular sets of relevant attributes. Thus a project which contains a number of different air compressors might have a table compiled specially for these. The attributes allocated to air compressors might include:

- component name   e.g. AC–F43B
- type   e.g Two-stage, air-cooled
- model   e.g. F43B
- supplier   e.g. Broom and Wade
- catalogue number   e.g. XJ45723
- rated output   e.g. 100 litres/m
- speed
- engine
- mounting
- overall weight
- initial cost
- hourly running cost
- maintenance supplier
- maintenance annual cost
- date required

Tables of component attributes can become the sources of bills of materials, component schedules of various kinds, and costings. The DBMS can select and extract the data required for a particular application, and arrange the information in a suitable format of schedule or report. Buildings often require schedules of components like doors, windows, cladding units, lintels, structural steel members, precast concrete units, sanitary appliances, water tanks, air conditioning ducts, light fittings, manholes and furniture.

Attributes of design units

Of course a CAD component is a design unit. It can be useful however to attach attributes to other elements in the design hierarchy. So a design unit can be any sub-division of the project design which has usefulness to the designer. An obvious unit within buildings is the room. Attributes attached to room spaces could be:

- name   e.g. Alexander Ward
- activity   e.g. ward
- floor area   e.g. 84 m^2
- floor surface   e.g. wood block
- ceiling finish
- wall surface area
- wall finish
- budget for furniture
- design temperature
- target opening date

This can be the information source for room use schedules. With calculations and sortations carried on surface areas, it is also possible to produce a schedule of finishes. Linking unit costs of various finishes can produce cost breakdowns of the various finishes.

Other types of design units which might have attributes attached for particular purposes include excavations, structural frames, service cores, service ducts, fire compartments and roof structures.

Attributes of drawings

Section 10.6 covered the definition of drawings. The attributes of each drawing such as a floor plan could include:

- project name,
- CAD model name,
- floor level,
- two sets of co-ordinates to specify the extent,
- component classes and design layers included in the drawing.

Such attributes might be maintained automatically by some CAD systems. Other attributes that could be useful are: model revision code; date of drawing extraction; purpose of drawing or trade.

Attributes of drawing sheets

Each drawing sheet plotted by the system could have attributes attached too, including for example:

- the sheet number,
- date,
- sheet title,
- list of drawings included on the sheet,
- design discipline,
- person responsible,
- person approving the sheet,
  etc.

The complete file of sheet information becomes a form of drawing sheet register for the project. The subject of such registers will be discussed in more detail in Chapter 12.

11.6 LINKING GRAPHICAL AND NON-GRAPHICAL DATA

In CAD work, an in-built linkage between the graphical and relational databases is very helpful to successful use.

If we look at the simple example shown in Table 11.1, the source of much of the data in File A would be the assembled graphical database for the project. This is because the numbers of instances of named components occurring within defined rooms in the hospital are built up automatically by the system as the user assembles the design.

The unit cost of each type of furniture in File B is a non-graphic attribute which would be input and stored – perhaps when the graphical component is created. Of course attributes like the unit cost could be updated later as the design evolves.

We see therefore that as the design work progresses, the graphical and non-graphical data can build up in unison.

Here I wish to highlight the way in which this graphical and non-graphical information can be related to one-another. From this dual source it is possible to generate a variety of useful reports and schedules, and perhaps to undertake costing studies for a project.

Fig. 11.1 shows the various constituents. For convenience, I have associated the term *library* with component graphics, and the term *database* with non-graphical information. This last is input to and held within a database management system. Each 'level' of the process as indicated in Fig. 11.1 will be considered in turn.

Level A: master library and master database

The design office as a whole will probably have established a master library and a master database containing information for all the potentially-useful components, design units, standard details, etc. This is corporate information rather than project-related.

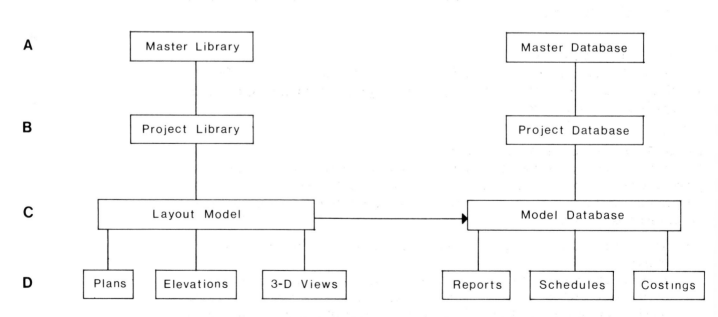

Fig 11.1 Relationship between graphical and non-graphic data.

The master library contains the graphical representations. This is one source of the graphical images, including annotation, that will appear on drawings. The master library is not directly used for projects, but rather copies of components are made and these are transferred for use on individual projects.

The master database contains all the non-graphic attributes and descriptions relating to the components, design units, etc. This too is corporate data.

Level B: project library and project database

A project library and project database will be compiled to hold the graphical/geometrical data and attributes respectively of all the components and design units relating to a specific project.

These will be derived from the master files, perhaps also from external sources such as RIBA Data Services. A proportion of the project data will of course be created specially for the individual project.

Level C: layout model

Using the graphics modelling system, the layout model is assembled by placement of instances of the project components. Parts of the layout model may also be compiled from design sub-models – from various sources – suitably adapted for the current circumstances (Fig. 10.1).

The information in the assembled layout model will provide the precise location, as well as classification information such as the design layer of each component instance.

Level C: model database

At any time in the design process it is possible for the computer to generate a version of the model database. This is a 'snapshot' taken to link:

- all the instances of the components currently assembled in the layout model (and their locations within the project, their class and design layer)   *with*
- the related attributes of these component instances. These would be available in the project database.

Level D: the drawing sheets

We are now familiar with the way in which drawing sheets can be generated from the assembled layout model. The sheets might contain plans, elevations, sections, and perhaps 3–D drawings of the model.

Level D: reports, schedules, costings

Earlier in this chapter, I explained how a database management system can be used to generate either routine or *ad hoc* reports and schedules, or deal with one-off queries.

Operating on the model database, the database management system can answer queries such as for example:

- list all the air-conditioning duct components, by level, in each building of the project;
- schedule and summate the costs of the light fittings in each warehouse zone;
- calculate the total cost of all the services elements on level 2 in the design model as it existed on two specified dates. The object of this would be to show the savings resulting from certain design changes.

```
* * * * * * * * * * * * * * * * * * * * * * * * * * * * * * * * * * * *
* * * * * * * * * * * * S T Y L E   S T O R E S   P L C * * * * * * * *
```

STORE   : BEVERLEY
AREA (SQ.FT.) : 2076.68

CODE	DESCRIPTION	SUPPLIER	SIZE	COST	LIN.FT.	SQ.FT.	QTY	TOT.COST	TOT.LF	TOT.SQ.FT
F1	GEN HANGING(4 SIDED)	USEFUL UNITS LTD	1400X1400	350.00	12.00	220.00	9	3150.00	108.00	1980.00
F3	GEN HANGING(2 SIDED)	USEFUL UNITS LTD	550 X 1150	325.00	5.00	300.00	3	975.00	15.00	900.00
F3A	GEN HANGING(2 SIDED)	USEFUL UNITS LTD	550 X 1150	325.00	5.00	300.00	3	975.00	15.00	900.00
W10	GEN MERCH HANGING	USEFUL UNITS LTD	1100 X 400	275.00	8.00	440.00	3	825.00	24.00	1320.00
W11	GEN MERCH HANGING	USEFUL UNITS LTD	1100 X 400	175.00	8.00	440.00	3	525.00	24.00	1320.00
W4	FORMAL SHOES - EM	USEFUL UNITS LTD	1100 X 400	200.00	8.00	440.00	2	400.00	16.00	880.00
W4A	FORMAL SHOES - RM	USEFUL UNITS LTD	1100 X 400	250.00	8.00	440.00	3	750.00	24.00	1320.00
W6	TRSRS/JEANS/YTHS/SPW	USEFUL UNITS LTD	1100 X 400	300.00	8.00	440.00	18	5400.00	144.00	7920.00
W6A	TRSRS/JEANS/YTHS/SPW	USEFUL UNITS LTD	550 X 400	150.00	4.00	220.00	2	300.00	8.00	440.00
W7	FOLDED MERCH	USEFUL UNITS LTD	1100 X 300	150.00	8.00	330.00	1	150.00	8.00	330.00
W8	GEN MERCH HANGING	USEFUL UNITS LTD	1100 X 400	225.00	8.00	440.00	4	900.00	32.00	1760.00
W9	GEN MERCH HANGING	USEFUL UNITS LTD	1100 X 400	250.00	8.00	440.00	3	750.00	24.00	1320.00
						GRAND TOTALS	54	15100.00	442.00	20390.00

Fig 11.2 Schedule produced from a Sonata model database.

When the CAD system is able to link the graphical and non-graphical data, then the model database becomes the depository for:

- component instances and other design units;
- numbers of occurrences, and their physical locations;
- classification information;
- non-graphic attributes.

During the progress of the design work, components will be added to the layout model. When required, a new version of the model database can be generated. This will automatically reflect the the revised numbers of occurrences and physical locations of components in the current model, as will all the schedules or reports generated from the model database.

Fig. 11.2 is an example of part of a schedule produced from a model database.

11.7 ATTRIBUTES NOT LINKED TO THE GRAPHICAL DATA

When the CAD system in use cannot cope with attribute information, then it is still possible to work with non-graphic data. But it must be done separately from the CAD work. A database management system, or perhaps a spreadsheet program is required. The work may be done either using this type of software running in the CAD computer, or else in any other machine such as an office micro.

Tables of attribute information are built up as illustrated in Section 11.5. However no data link can be established between these attributes and the graphical layout model. This means that information like the number of component instances, their locations and classification would have to be input manually to the non-graphic attribute tables. For example in the hospital example, the numbers of instances of named furniture elements occurring in each room space would have to be re-input to the tables. Likewise, when the design is revised, the user is responsible for making all the necessary alterations by hand.

CHAPTER 12

Project Control and Quality Assurance

Control procedures are reviewed in this chapter. These can be applied both to assist in the planning of projects and in the monitoring of their progress. It is by these means that the quality of the design work can be more assured. Among the procedures are design cost control and the maintenance of registers of design documents (i.e. drawing sheets, schedules, etc.). These document registers can, if required, be extended to provide a simple means of monitoring job work-in-progress.

Arguably the most important procedures are those to control the checking, approval and release of design information to others. The means for controlling design revisions must be inherent in these. This whole subject is comprehensively covered in relation to computer modelling.

Eventually design information will be communicated electronically, but in the meantime there continues to be much exchanging of documents on paper. A procedure for recording such exchanges is discussed. Many of the foregoing ideas are combined into the concept of a history file for computer models and documents. This can provide a comprehensive audit on the design work.

The chapter ends with a brief discussion on quality assurance and its influence on design work.

12.1 INTRODUCTION

I have dealt with the topic of project management and planning in Chapter 8. Project control relates to the on-going monitoring of the design work-in-progress so as to detect deviations either from the plan or from budgeted costs. When these are detected and if they are serious, then either action must be taken to bring the project back on course, or else an amended plan must be drawn up. There are various facets to project control and a number of administrative systems can be established to assist.

Every design office already has some control procedures. These are either in full operation, or have been proposed by well-intentioned people but operate in a patchy fashion. Every office functions in different ways. With the introduction of CAD, it is not recommended that effective systems are replaced with the various suggestions that follow. Rather, the strong aspects of existing systems should be maintained so that disruption is minimised. However, it is probable that the office could benefit if existing control is tightened or extended. This can be done if the strong points of existing procedures can evolve, and the discussion in this chapter may provide some pointers on how this can happen. Remember that CAD must coexist with manual design and this must be reflected in all procedures implemented.

12.2 DESIGN COSTS

With regard to design costs, these are normally monitored using time sheets. The staff hours worked are allocated to numbered projects. Various expenses incurred may also be included. Sometimes timesheets are monitored with the aid of a cost control computer program. The details of the procedures vary somewhat in different design offices.

With the introduction of CAD, the existing procedures for design cost control do not have to be altered, apart that is from the monitoring and inclusion of the costs of CAD resources. These can be monitored using a workstation logging procedure, and this topic was discussed in Section 6.12.

12.3 REGISTERS FOR DESIGN DOCUMENTS

The tangible forms of output from a CAD system are the drawing sheets, schedules and reports. It is essential to maintain some form of record of these documents.

Document registers can be produced by hand in the office as they always have been.

Alternatively the office could set up its own computer system for document registers. This could use a database management program, possibly running on a microcomputer. I shall explain how this can be done. Refer also to Chapter 11 where the use of database management systems for manipulating non-graphic design information was discussed. In particular see Section 11.5 where I showed how attributes could be attached to drawing sheets.

Document attributes

Each document would be described in one record (or line) in a table. The fields (or columns) of the table would be the attributes or descriptors of the documents. Obviously one attribute would be the document reference or name. In Section 10.7, I covered the subject of drawing sheet names. Documents like schedules and specifications could be referenced along similar lines. A simple document register system might use the following attributes:

- Document Reference. e.g. including (as discussed in Section 10.7):
    project reference;
    computer model number (for CAD documents);
    control status (see Section 12.5);
    revision code;
    design discipline;
    sheet number.
- Document title.
- Initials (or name) of person responsible.
- Release Date.

The complete table would include all the documents produced which are associated with one project. A rather more ambitious scheme would include all the documents produced in the whole design office.

From such a computer-based table of information, pre-formatted reports could be produced at intervals using the facilities of a database program. It could also cope with *ad hoc* queries and reports. For example we could display on a computer screen all the drawings produced between two dates by one named individual in the drawing office.

Retrieval of design information

In practice there is really no limit to the number of fields or attributes that can describe each document. There must of course be a purpose for each, and staff must be motivated to maintain the information. An obvious extension is to code the document content, so that the register system can be used for information retrieval purposes. Document content might be by location within the project (e.g. Block B, 4th Floor), by trade or building element, or by components included. The European CISfB system of classification or other suitable scheme might be employed.

Note that the document register must cope both with manually-produced and computer-produced documents, since these will coexist for a long time.

12.4 MONITORING DESIGN WORK-IN-PROGRESS

Let us now look at an application that goes somewhat beyond the simple maintenance of document registers. With a database management system it is a simple process to monitor the status of the work-in-progress.

A viable project plan for a small job might be based merely on a register of documents. An initial register would be compiled at the start of the job, as a part of the project planning (see Section 8.2). Each document would be treated as one task. Then a member of staff would be assigned to produce each document, starting and finishing the task within specified dates. Project control is the monitoring of whether the documents are in fact being completed by these dates.

Remember that one record in a table can be assigned for each document. Fields of the table could be assigned for:

- Document reference, e.g. including:
    - project reference;
    - computer model number (for CAD documents);
    - control status;
    - revision code;
    - design discipline;
    - sheet number.
- Document title.
- Initials (or name) of person responsible.
- Scheduled start date.
- Scheduled finish date.
- Estimated manhours required.
- Percentage of task complete at current date.
- Release Date.

If estimated manhours and percentage completions relating to individual documents are indeed maintained as suggested above, then we have the makings of a simple system for drawing office control. The records should be updated at frequent intervals.

A simple mathematical treatment of the entire document register for a project could be undertaken automatically with the computer. This would produce an estimate of the percentage completion of the whole project at the current date. Likewise the manhours needed to complete the entire project could be calculated. The results of successive analyses could be plotted as graphs (normally S-curves) linking percentage completion with date.

This makes the not unreasonable assumption that much design project work can be related to the tasks involved in producing the design documents.

PHASE

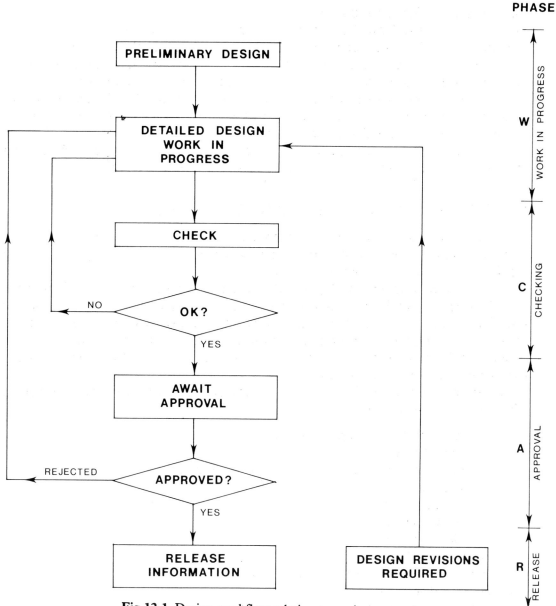

Fig 12.1 Design workflow relating to project control.

12.5 CHECKING, APPROVALS AND RELEASE CONTROL

Design workflow and project control phases

Fig. 12.1 shows the concept of design workflow which emphasises the project control aspects. Four phases are indicated.

W *Work-in-progress*

   The essential features are that any parcel of design work progresses within the design office, and during this time it is obviously in a state of flux. This is the 'work-in-progress' phase. At some stage this work is judged to be complete or is halted. The work then moves on to the 'checking' phase.

C *Checking*

In practice it is the design documentation that must be checked now. If errors are found then the project must revert to the 'work-in-progress' phase. Otherwise the 'approval' phase may be entered.

A *Approval*

The documentation is now approved or rejected. If approved the documentation can be released, otherwise the project must revert to the 'work-in-progress' phase.

R *Release*

Here the documentation is released for use by others, such as the client, other members of the design team, for the tendering procedure or for construction purposes. If at any time a revision to the design becomes necessary, then the project must revert back to the 'work-in-progress' phase.

This concept of work flow, with the pattern of four control phases is already operated in most design offices. The 'tightness' or otherwise of the checking and approval procedures does tend to vary.

Now let us examine how this pattern can be applied to CAD work. Control is especially important because a version of the design information is maintained within the computer in an intangible form. We have to know how physical documents relate to this computer data.

Fig. 12.2 sets out how the control phases already discussed can be related to the CAD model and the documents. According to this scheme, computer models and the documents are always provided with a 'Control status' which conforms with the relevant control phase, viz:

Code	Control Status
W	Work-in-progress on the design
C	Checking
A	Awaiting approval
R	Released information

As discussed in Section 10.7, the status code could be included in the unique document reference and marked on the document itself. Then the status is clear to all.

Essentially the design 'work-in-progress' phase includes the model building process. During this phase of the work, plots may be ordered. These are 'working drawings' in the sense that they are office copies for use by the designers themselves. They are marked with control status 'W' which indicates that they are not intended for use by others. If indeed they are passed to others, the recipients must be given to understand in unmistakable terms that the drawings are subject to alteration without notice.

Locking of model

In Fig. 12.2, the additional concept introduced is the 'locking' and 'unlocking' of the CAD model.

An unlocked model can be freely worked on – by adding, modifying and deleting information – without serious concern for the effect on others.

A locked model must not undergo any changes whatsoever.

The act of physically locking a computer model is done by a responsible person, such as the CAD co-ordinator, the computer manager or a project manager. Locking the

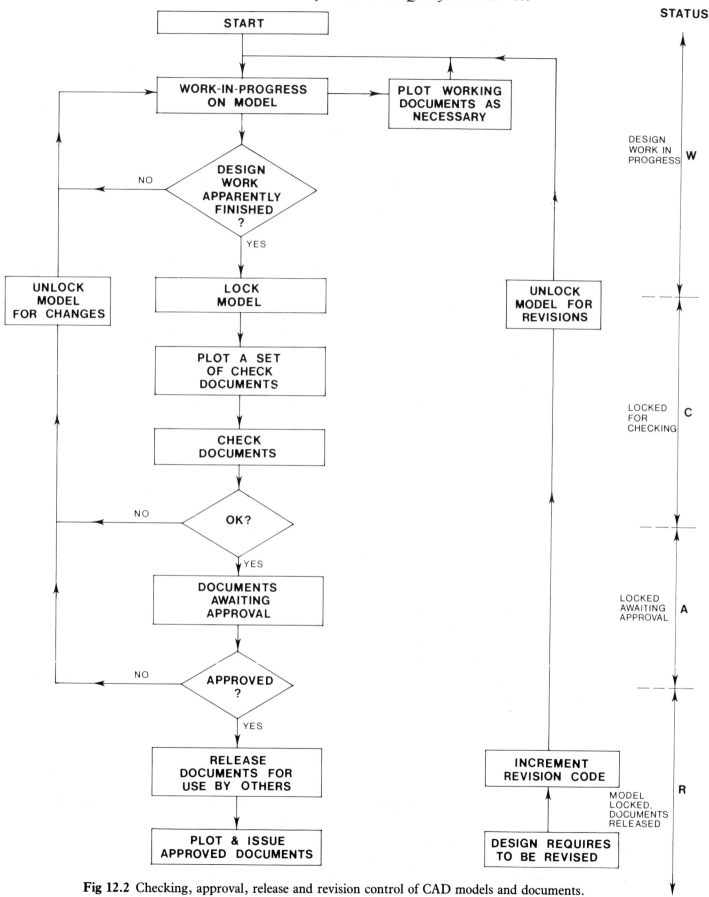

Fig 12.2 Checking, approval, release and revision control of CAD models and documents.

computer file which contains the model should involve the following, (where the computer's operating system permits):

- Changing the file access mode from 'read-write-delete' capability to 'read-only' capability. This allows operators to view the model, and to produce drawings – but it is impossible to change anything while set thus.
- Passwords would be set in such a fashion that only the responsible person can alter the file access mode of this model. This ensures that it is impossible for others to change the model.
- The model name should be changed to reflect the change in control status. This involves changing the status code letter from 'W' to 'C' (for checking).

The act of unlocking a model can only be done by the responsible person. It requires a knowledge of the relevant password, and changing the file access back to 'read-write-delete' status. The control status code letter is changed back to 'W' to indicate that the model is back in the 'work-in-progress' status and the revision letter is automatically altered if appropriate. Now the operators concerned can continue to modify the model as required.

Checking documents

Since design information is usually communicated by issuing drawings or other documents, it is these documents which essentially must be checked. When computer models are communicated directly, then it is the model itself that must be checked. The same principles applies to approvals.

After the model has been locked, then a set of drawings are plotted. These are the 'check-drawings'. They are sent to the independent checker who, depending on requirements, might be in another department or another firm. It is important that the responsibility is shouldered by a named individual and organisation. Items checked can be highlighted in yellow while errors or required alterations can be marked clearly on the documents with a red felt pen.

With manually-produced documents, the checker could sign the master document itself. With CAD documents, either the plotted master itself can be signed, or on a separate control document the signature can be set against a unique reference to the checked document. A positive means for establishing the unique reference is for the computer system to automatically print or plot on each document the document name, and the date and precise time (to the minute) of the document production.

If the checker is satisfied that all is well with the set of check-drawings produced from the model, then the responsible person referred to above should move the model on the 'approved' status. The model remains locked, but the status code of the model is changed from 'C' to 'A'. Then everyone knows what has happened. If corrections are necessary, the status reverts to 'W' and the model is unlocked.

Approvals of documents

A set of documents can be plotted for the approver. This is usually a senior person. The document names, marked with the status letter 'A' and the date and time of production will appear on these approval documents. If approved these are signed directly by the approver, or else the separate control document is signed. The model is 'released'.

Released status

The model remains locked and so has 'read-only' access, which means that documents can be produced from it but it cannot be altered. The model is renamed with the 'R' status and documents produced thereafter also contain this code in their name. This indicates that the documents are releasable for use by others.

This is always an appropriate time for archiving the model on magnetic tape. If required, documents may be microfilmed too.

12.6 CONTROL OF DOCUMENT EXCHANGES

Maintaining a control over the document exchanges with other organisations is usually a major administrative headache in most design offices. The word 'exchange' is used here because in practice documents are issued, and received from elsewhere.

Again, a manual system for keeping track of document exchanges could be instituted in the office. But a computer database system could be set up to tackle this job. This might be along the following lines.

Each document issued or received would be described in one record (or line) in a table. Fields (or columns) within these records could be assigned for:

- Document Reference. e.g. including:
     project reference;
     computer model number (for CAD documents);
     control status;
     revision code;
     design discipline;
     sheet number.
- Document title.
- Initials (or name) of person responsible.
- Release date.
- Date of document exchange.
- Staff member responsible in this office.
- Staff member responsible in outside organisation.
- Number of copies sent/received.
- Size of drawing print or printed document.
- Scale of drawing.

As in the implementation of document registers, the database system should allow pre-formatted reports to be produced when required. Also it ought to cope with *ad hoc* queries and reports regarding exchanges of documents. For example, all the documents sent to a particular firm between any two dates could be listed.

Planning of document exchanges

A new record could be added when each exchange is actually made. It is preferable, however, if the exchanges are planned beforehand and records inserted then. Some of the fields like 'Date of document exchange' and 'Number of copies' could be left blank initially, and filled in later on to note that the transfer has actually been made.

Alternatively if target dates are inserted at the project planning stage, the system could be easily configured to remind staff to make the exchanges at the appropriate dates.

With the use of a suitable relational database system, the Document Register and the Document Exchange records could be combined. This would save the re-entry of some of the details of individual documents. The common key in the two files would be the unique reference of each document.

12.7 DESIGN REVISION CONTROL

Design changes present few problems early in the design process, or where just a few designers working in one office are concerned.

After documents have been released and issued to others, there must be greater concern. At this time they are not mere 'design changes' but are 'revisions'. This reflects the fact that the control procedure must be invoked.

A revision control system is of course necessary whether or not CAD is being applied. Where drawings are produced manually, the normal procedure is for the drawing master to be modified and a note of the revision is added to a revision box. Provided this is done it is clear that the revision has been carried out. The details, including the date and the person responsible, remain in the revision box. This is the drawing's 'audit trail'. It is of course very important that copies of the revised master are issued to all parties that need the revised information.

With CAD modelling, it may or may not be easy to change the computer model. After the model has been revised however, it is a simple matter to order up a new plot. This will be a clean new master document which must replace the previous master – so the procedure is different from manual methods.

It is time-wasting to keep on making minor revisions, so efforts should be made to batch several modifications together into one revision.

Revision procedure

When a revision becomes necessary to a 'released' design, the procedure is indicated in Fig. 12.2 and it involves the following:

(1) The revision code is incremented. Sometimes the first released version has no revision letter, and so now revision 'A' must be set. Revision 'C' changes to 'D'.

(2) The model is unlocked, and its control status changed from 'R' to 'W'. The model is revised.

A potential problem is that the effects of a revision to the model might well be reflected on several documents. There is the danger then that new plots will in the event be made of some, but not all, of these. Some will be reissued later, while others will not. This is one way in which issued drawings can easily get out of step with the model.

The person revising the model must carefully assess the effects of the revision, and determine which documents will change. If there is any doubt whatsoever, then plans must be made to reproduce the whole document set.

(3) The path must then lie through the checking, approval and release phases again. When a partial set of revised documents are produced, the checker must ensure that these and no others would be affected by the model revision.

(4) After the new release, careful consideration must be given to reissue to all parties that require the revised information.

No manual alterations should ever be made to a plotted master drawing as a matter of expediency, while omitting to make the equivalent changes to the model. This is, of course, another way in which the computer model can get out of phase with the drawings.

The basic problem is that the only true 'master' is the computer model. Revisions to this must be controlled, and care taken that all documents in current use are derived from or relate to the latest version of the model.

It remains an important cross-check that each CAD-plotted document must itself be marked in some way. The well-established practice for drawings, of including the revision box and appending a revision-letter to the drawing reference, should be maintained. The CAD operator making changes must be responsible for seeing that this is done.

The best defence against errors or misunderstandings is to institute office rules which restrict particular operators to manipulations involving only the work of their own discipline. In a multi-discipline environment, they can be limited to operations on those components that lie within a restricted range of categories or layers of design information.

12.8 DOCUMENT HISTORY FILE AND AUDIT TRAILS

Perhaps it is now clear that with the help of a database management system, it is possible to maintain a file which contains all the records appertaining to models and documents. This is a consolidation of:

(1) The information that would ordinarily be kept in the document register (Section 12.3);
(2) Information on work-in-progress (Section 12.4);
(3) Dates and persons responsible for moving computer models through the various control phases, i.e. W, C, A and R;
(4) Records of documents plotted or printed;
(5) Information relating to revisions, e.g. dates; person responsible for making the revision; nature of revision;
(6) Information on document exchanges.

The database management system (DBMS) would be able to support interrogation of the history file, and to produce reports as required. Audit trails showing the progress of project design work could be produced.

This high level of recording and control would of course be virtually impossible without a computer DBMS. Where the DBMS is a separate identity from the CAD system, the manual input of information might still prove to be a little daunting. It could be done for the large or critical projects where good control is particularly important. In other cases, some of the elements of the control procedure discussed could be instituted and others not, according to management preferences.

Perhaps we should look to the time when the CAD system itself will contain all the elements of such control procedures – for then much of the recording could be entirely automatic.

12.9 QUALITY ASSURANCE AND DESIGN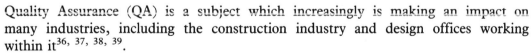

Quality Assurance (QA) is a subject which increasingly is making an impact on many industries, including the construction industry and design offices working within it[36, 37, 38, 39].

The scope of QA relates to all activities, functions and procedures which are concerned with the attainment of quality. This means the fitness of purpose of the service, product or project. In construction design, it implies the application of formal management techniques, the purposes of which are:

(1) to increase the probability that a project being designed will be fit for its purpose when completed;
(2) to assure the client, and possibly the users of the project and others, that quality can and will be achieved in the project.

Formal procedures include written descriptions of activities, and inspections and independent audits to check that these activities are being correctly carried out.

In construction design, QA can be important to assure:

- reliability,
- public safety,
- prudent expenditure, particularly of public funds,
- completion of the project on time.

QA in design is needed in an industry which produces unique projects which cannot easily be tested in actual use. The decision to introduce QA must be taken by, and have the commitment of, senior management of the organisation. It should be adopted when the advantages are judged to outweigh the disadvantages. One advantage is the satisfaction of producing better design work, as well as the good client and public relations – and marketing advantage – that stem from this. Fewer of the costly problems should appear late in the design process or during the construction phase. A disadvantage of QA is that some additional cost may be involved.

QA has been adopted for large and complex projects like power stations and petrochemical plants. It seems likely that large client organisations including public authorities will press for QA to be applied on progressively smaller construction projects in the future.

A *Quality System* is the management structure, responsibilities and *Quality Procedures* which a design office has to set up.

A *Quality Programme* is a written description of a Quality System. It applies to the work of the whole, or a part, of the *organisation generally* and not to a specific project.

A *Quality Plan* is a document derived from the Quality Programme, extended if necessary, which relates to a *particular project* on which QA is to be adopted.

These and other definitions are set out in the references.

With regard to the use of CAD for project design, a Quality System would have to include a management structure for the CAD implementation, as well as the responsibilities and procedures adopted for the planning and control of the CAD work. Procedures for project management, for administering document registers, for checking, approvals, revision control, and recording of document exchanges and audit trails – as discussed in this chapter – are all relevant. QA means that not only do such procedures have to be proposed, they must be defined in writing and proved in operation. Thus a Quality Programme and a Quality Plan for a specific project would have to include these in written form.

QA is much concerned with assigning responsibilities to individuals. This is why I have stressed throughout this book the various roles and responsibilities. This is why, for example, the names of individuals responsible for checking, approving and releasing information should all be carefully recorded as part of the control procedure.

CHAPTER 13

Construction, Space Planning and Building Management

Previous chapters have dealt with various facets of the application of CAD systems to the design of new projects. Here we switch to take a look at CAD applications in the planning of construction work, and in the space planning and management of buildings that already exist.

The techniques covered will include a look at specialist applications software for strategic space planning. They will also extend from simple drafting techniques as used for straightforward space layout, to more sophisticated modelling, visualisation of interiors and facilities management using database software tools.

The CAD drawings and models built up within the design phase of a project can become an asset which is potentially valuable to others in the future. Those that might benefit are the constructors, clients, interior designers, office or factory managers, those responsible for maintenance and operation of the facilities, and perhaps the designers subsequently charged with the design of refurbishment works.

Indeed the original designers ought to be more aware of diverse applications because opportunities can occur for them to expand their own role.

13.1 CONSTRUCTION PLANNING

The drawings and schedules produced by the design professionals rarely form the full complement of documents from which a project can be constructed by site operatives. After the main contractor has been appointed, it is common practice for a variety of new drawings and schedules to be made. These will include further design details, shop drawings, construction proposals, temporary works details, as well as further extractions of lists and schedules for procurement purposes.

It is too much to expect that the designer's CAD model or drawings can be the source of all this additional documentation. However the CAD system is certainly a source of some of it, including much of the dimensional framework for the project, of additional measurement required for construction purposes, of spatial layout, and outlines of the permanent works. In essence, it could be the information backdrop on which much of the additional construction documentation could be hung.

Sometimes there are practical problems, and perhaps some legal constraints too, for a CAD model may contain more (or less) information as compared with the tender and construction drawings which are recognised as legal documents.

These ought not to be serious problems that cannot be overcome. Indeed there are likely to be increasingly strong pressures to overcome them as has been the case in other industries. In particular, motor-car manufacturers throughout the world have been applying strong pressures on their own components suppliers. They are requiring them not only to supply the mechanical parts, but also the design data from the suppliers' CAD systems in a format that is fully compatible with their own CAD systems.

Let us now look at a few possibilities. Using additional design layers (see Section 10.5) the contractor could add temporary works, specify pipe runs more clearly, produce larger scale drawings of zones and then add more detail. He could annotate components with suppliers' names and order codes, define construction phases, mark elements that have been completed, and add site features not already shown. It would be possible to place the site accommodation, plan locations for stockpiles, mark the extent of excavations, check out the site access routes, set out and measure the layout of security fencing and mark adjacent property. Contractors always need a host of measurements which are never shown on the drawings supplied to them. These are almost random thoughts but perhaps will provide some ideas.

3–D models could prove particularly valuable. Consider for example the interaction of two or more tower cranes on a congested site. These have to be able to pick up materials either from parked delivery vehicles or from site stocks, and supply them to a multitude of locations and levels. The cranes (and cables) must not come into contact with existing or adjacent construction, temporary works, or each other!

13.2 A NEW SERVICE PROVIDED BY THE ORIGINAL DESIGNERS

When the designers operate in a different organisation to that of the constructors, as is common, they may be reluctant to hand over valuable information to others.

The multiplicity of incompatible CAD systems is a potential problem too. Sometimes several design practices contribute towards the design of a project, and perhaps more than one CAD system was employed. The contractor's staff may not be familiar with the operation of, or have any form of access to, the same type of CAD system. The problems involved in transferring data from one system to another were aired in Section 6.10. These are not insurmountable, but they add to the difficulties.

One solution would be for the designer's office to gear itself to providing an extra service to the contractors and other interested parties. This would involve supplying under an agreement:

- Copies of the computer design models.
- Loan of one or more CAD workstations – either self-contained stations or terminals connected (perhaps through telecommunication links) with the designer's own system.
- Loan of experienced and project-knowledgable operators.

The company making use of the service would be at liberty to add addition material to the model, or even with care to change what is already there – at its own risk. Obviously to a contractor building the project, the design information supplied under the contract would still be binding as far as the construction work is concerned.

13.3 STRATEGIC SPACE PLANNING FOR BUSINESS CONCERNS

This section reviews some computer application techniques which have been developed for strategic space planning. These relate to the manner in which business hierarchies can be fitted into and occupy floor space. The hierarchies include departments, functional work groups, functional work spaces (such as store rooms) and individual staff. The strategic planning might be carried out either:

(a) by the client and architect during the conceptual design phase of a new building; *or*

(b) by business management, perhaps with its advisers, before a physical move is made to another building, or when existing accommodation is going to be rearranged or refurbished.

The purpose is to find how the business hierarchies can be best arranged within building space that either already exists, or for which a conceptual plan exists. The arrangement must have regard to the functions and activities, space needs and physical communications between people. The aim is to achieve greater efficiency in operations, to improve an organisation's flow of work, to obtain staff time savings, improved communications, and a more accurate reporting mechanism with better data for decision-making by management.

First some data has to be gathered on the organisation's structure – its breakdown into departments, groups or even individuals – as well as the nature of activities carried out. Next the space available for allocation is required, in terms of buildings, floors and areas within floors. Details of existing occupancy may be needed.

For the initial planning, the activities or groups of people are listed. Then these are assigned to buildings and distributed to floors, this assignment being done using subjective judgement. The likely interactions are then specified. This can can be done using bubble diagrams or matrices. The size of bubbles represent the area required and the connecting lines show interactions. Matrices connect the activities or groups in tabular form and indicate grades of interaction such as high, low or prohibited. For example there might be high interaction between the reception foyer and the conference room, but prohibited connection between a customer circulation area and the room housing the company's mainframe computer.

Some space planning systems provide 'blocking and stacking' features. Stacking is a technique for showing diagrammatically the space – available and occupied – on each floor, and how the activities or groups fit.

Blocking is the technique whereby algorithms enable the computer to automatically assign positions to the organisation's departments or groups. These are indicated by coloured blocks on a floor planning grid, perhaps of one metre squares. The computer may also display the degree of interactions between these blocks using coloured lines, or lines of varying thickness. The aim is to get short, thick lines, although some long, thin ones may be permissible too. Of course, automatic blocking is no replacement for human judgement in these matters, but it provides the user with suggestions. He can interact with the system to try to improve the layout. A system may adopt a scoring system, where a score is calculated by the computer as being the product of interactions and distance. Clearly in this case the total score has to be minimised.

Strategic planning techiques like blocking and stacking are fairly popular in the United States. These are techniques that can augment a planning activity which basically involves much human judgement, yet requires some specialist expertise in its application. The techniques are therefore likely to appeal most to full-time planners operating in the large companies which have plenty of space to plan. They may also appeal to advisors to such companies, and to property or estate management concerns.

These methods are usefully applied for sophisticated 'what-if' investigations and in the assessment of alternative layout schemes. The justification for their use is based on the high asset value of building space, and the importance of effective layout in the operation of business enterprises.

13.4 SPACE LAYOUT AND FLOOR PLANS

In its simplest form, space layout planning with computer aids can be based merely on a computer drafting system.

Basically all that is needed is an outline of each floor area. This must show the positions of existing features like doors, windows, stairs, fixed services like radiators and power outlets, and fixed equipment like safes and large computers. Either this information can be digitised or scanned from architects' floor plans, or can be input directly from the measurements taken during a survey.

A computer-based library must be built up, containing the shapes of all the significant objects that must be accommodated. These may include desks, chairs, bookcases, filing cabinets, office machines, computer terminals, production machines, partitions, display units, lighting units, vehicles – according to the purpose of the building.

These shapes can be retrieved from the library one at a time, and arranged on the floor plan. They can be moved around in much the same fashion as we would have moved cardboard shapes on a floor plan drawing. Account is taken of the existing features, as well as the access and maintenance spaces required. The process continues until one or more suitable schemes are achieved. By this process, various layouts can be examined quickly and more conveniently compared with the traditional methods on paper. The 2–D layering facilities of the drafting system enable various classes of items to be placed on different layers. After schemes are finished, these layers of information can be combined into useful combinations for display.

By these means the space planner could study how the planned seating arrangements relate to the positions of hot-water radiators, how air-conditioning grilles relate to new partitions, and whether the office equipment will be within range of power outlets. A colour-graphics screen helps in the differentiation of the different layers or classes of information. By these means too, floor plans can be quickly produced on paper and at this level the process is both simple and straightforward to apply. Figs. 13.1 to 13.4 are examples of such floor plans.

13.5 VISUALISATION AND INTERIOR DESIGN

As the power of 3–D modelling and the quality of computer visualisation increases, so the benefits of adopting these techniques for space planning and interior design[40] are becoming better understood.

The efficiency of many activities are critically dependent on the nature of interior surroundings. Human behaviour is often influenced by the aesthetics of the environment. Sales and marketing people know this well, but such people inevitably have difficulty in appreciating how an planned interior will appear when the only evidence with which to judge it is a space planner's floor plan.

We can perhaps look for an example from the retailing environment. Store layouts have to be changed frequently, and changes always have to be effected in very short timescales. In these circumstances, planning for the future is a continuing activity. Bearing in mind the high value per unit area of certain retail floor space, it is clear that much ingenuity must go into ensuring that it is effectively utilised. However, the floor space is not the main interest, for various kinds of dispays must be mounted and customers' attention directed in various directions and levels.

These planners and the marketing people concerned are not likely to be satisfied with floor plans in line-drawing form. For lay persons, as opposed to skilled designers, these are no longer adequate when today it is possible to created 3–D views of proposals.

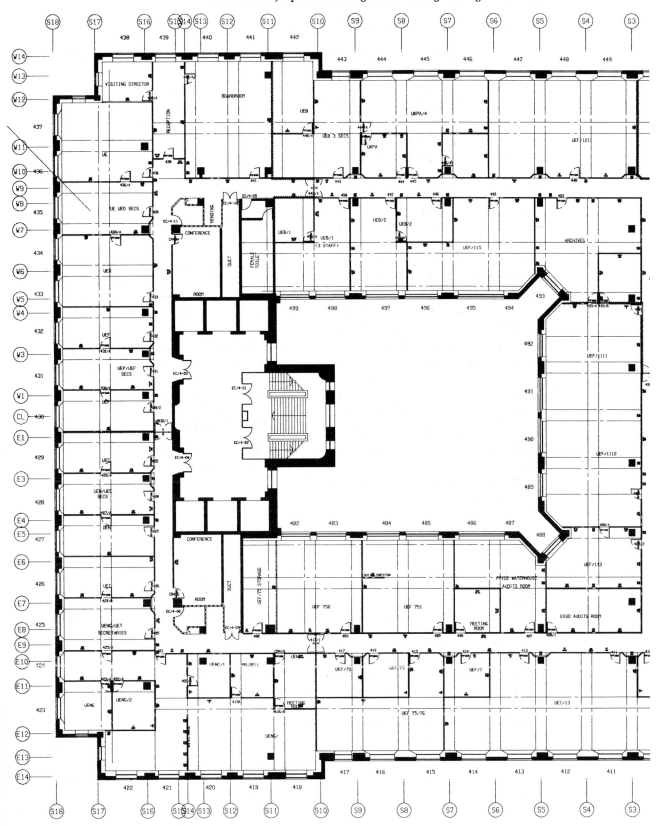

Fig 13.1 Shell Mex House Refurbishment. Partition layout. (*Courtesy*: GMW Partnership, London.)

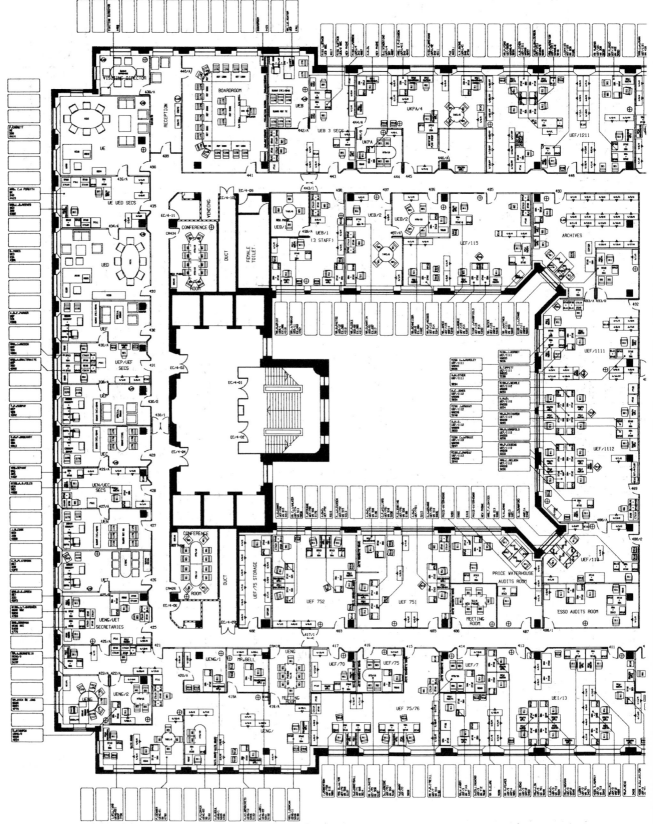

Fig 13.2 Shell Mex House Refurbishment. Furniture layout. (*Courtesy*: GMW Partnership, London.)

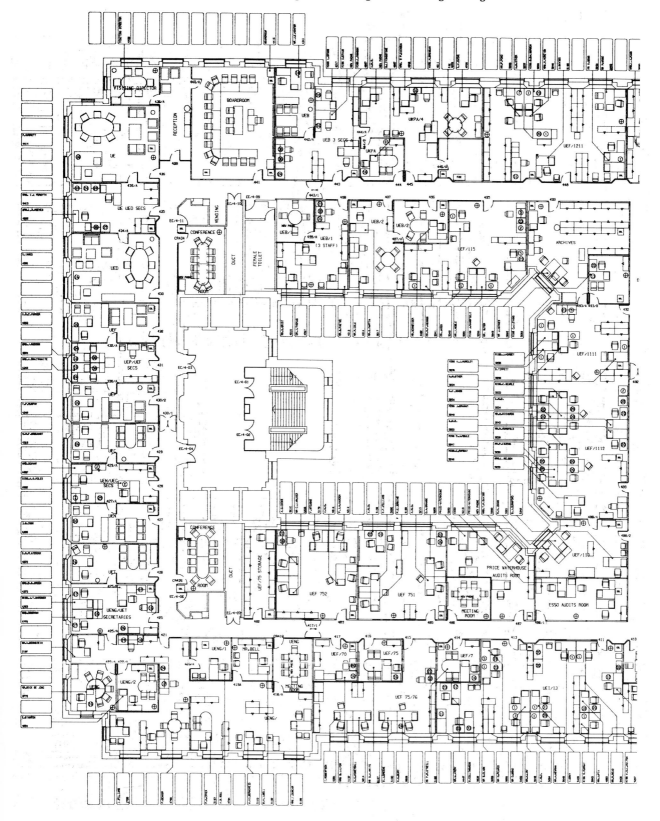

Fig 13.3 Shell Mex House Refurbishment. Computer equipment layout. (*Courtesy*: GMW Partnership, London.)

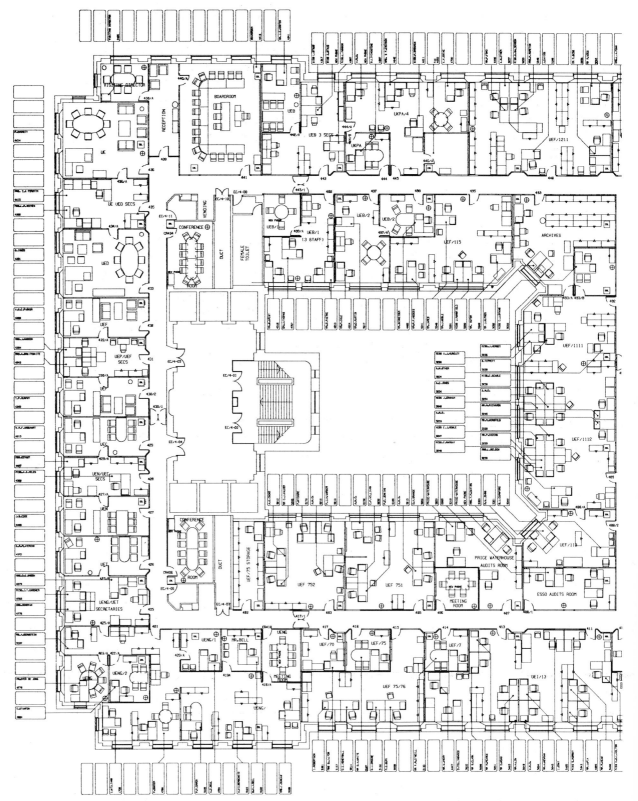

Fig 13.4 Shell Mex House Refurbishment. Telephone locations. (*Courtesy*: GMW Partnership, London.)

Figs. 13.5 and 13.6 are views of a building shell and the same area after fittings have been added.

Exquisite 3–D colour views can be produced for consideration. While viewing the screen, the positions of furniture, fittings, displays can be moved and changes can be made to the colour schemes too. Plate 6 is a computer-generated view of a theatre auditorium, while Plate 7 provides a realistic impression of how a design office will appear.

Displays, plots and screen photographs are easily produced. Other presentation techniques including audio-visual treatments and animations were discussed in Section 7.7.

13.6 FACILITIES MANAGEMENT AND SCHEDULING

Used in this context, 'facilities' is a general term to cover all the furniture and equipment which is provided within or around buildings. Examples are desks, chairs, display racks, movable partitions, telephones, facsimile machines, personal computers, typewriters, coffee-making machines, vacuum cleaners, lawn mowers. Certain fixed items within the fabric of the building might be included, with examples being electric power outlets, lighting fixtures, sanitary fittings, door locks and other security equipment, or indeed any elements of the engineering services. Facilities Management (FM) involves maintaining an inventory of all the facilities. It is essentially concerned with the effective utilisation of space and equipment. It can be applied:

- during conceptual design of a building,
- during detail design,
- for planning of furniture and equipment layout for the first occupancy,
- for subsequent changes of use/occupancy, including the re-siting of individuals, departments, equipment or production processes,
- for the planning of refurbishment work,
- also for the on-going reviewing of facilities generally.

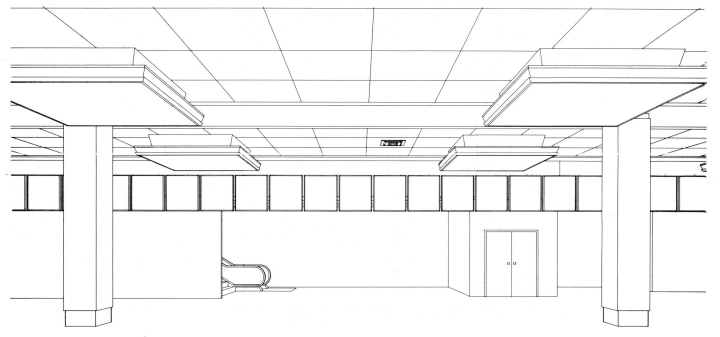

Fig 13.5 Retail interior – the shell. (*Courtesy*: Munro & Partners, Glasgow.)

Fig 13.6 Retail interior – Fitments. (*Courtesy*: Munro & Partners, Glasgow.)

FM can contribute to the general management of the business, to purchasing, stocking of spares, and overall maintenance and repairs.

There is no limit to the types of business environment covered, and so FM is used for offices, commercial and retail premises, hospitals, hotels, universities, defence establishments, process plants like oil refineries, offshore platforms, ships, warehouses and factories. With the last of these, the facilities covered might include the machine-tools and other manufacturing equipment.

FM is therefore the concern of property, estate, or office managers; space planners; interior designers; as well as architects, engineers, contractors, furniture suppliers and many others.

The key to practical FM lies in the space planning techniques already discussed in this chapter, combined with database management as discussed in Chapter 11. The computer has a role simply because managers soon become overwhelmed with the volume of records if they are maintained on paper, and cannot cope with continual changes. Provided a suitable classification scheme is instituted, the computer has the advantage that it can automatically print or plot selective reports which are clear, meaningful, and up-to-date.

Let us first of all deal with the graphics side of the operation. Where computer models or computer drawings were created in the project design phase, it is sensible if these can be retained and reused for on-going space planning and FM purposes. Facilities would be added, where necessary, to the computer model or drawings. For facilities classification, it normally would be appropriate to adopt an additional batch of design layers. In this way, each facility can be tied in to a location within the project, and new views or drawings can be plotted at any time to highlight particular groups of facilities.

Where the facility managers do not have access directly to suitable computer equipment, or do not have the CAD skills or experience, then business arrangements might be made with the original designers as discussed in Section 13.3. On the other hand, several organisations having a large volume of facilities management work have been investing in CAD systems which they commit wholly to this application. Such organisations include retail chains, manufacturing concerns and developers.

In many cases, the main emphasis of FM is not on the graphics side but relates to non-graphic attributes and descriptions of facilities. The techniques were covered in Chapter 11. Many facilities managers will wish to record attributes of equipment such as:

- name of facility,
- description,
- supplier: name, address, telephone number, account number,
- model, size, rating, colour, material,
- parts/catalogue number,
- location, e.g. link to CAD model database or drawing,
- department/person responsible,
- supplier's warranty status,
- date of installation,
- renewal date (if applicable),
- details of last inspection, e.g. date, who undertook it and condition,
- details of next inspection, e.g. date, who will undertake it,
- spares: location and number available,
- cost of facility,
- book value, i.e. original cost less depreciation,
- paper files, i.e. who keeps them and where.

A database system with alphanumeric display screens only can be used to maintain all such information. If necessary, security in the form of password control can be applied to some of the attributes, so that access is limited to certain managers.

The strength of such a system is two-fold:

(1) Ability to query any facility record at any time.
(2) The records can be sorted according to any of the attributes. Then *ad hoc* or customised reports can be printed to assist in area analysis, for estimating, cost comparisons, ordering purposes, asset management, and especially for maintenance scheduling.

Examples of the reports that might be produced are:

- Summarise all the facilities with asset values.
- Print a list of all personal computers, with suppliers.
- Print a list of all internal telephone numbers.
- Print a list of all preventive maintenance checks to be carried out within the next month.
- Detail all the equipment maintenance contracts during last (next) year.
- Schedule all outstanding repairs.
- Itemise all repairs carried out to a particular set of elevators during the last two years.

These examples are perhaps adequate to demonstrate that a FM system has the potential to achieve greater efficiency in operations, to improve an organisation's flow of work, to obtain staff time savings, improved communications, and provide a more accurate reporting mechanism with better data for decision-making by management.

In practice, there must be an emphasis on simplicity of operation perhaps with screen menus, help facilities, and assistance with customisation for individual circumstances. Special routines to assist in checking of input are important in the interests of maintaining correct information.

CHAPTER 14

Concluding Remarks and The Future

Only a tiny, well-funded, design office can make a sudden leap to embrace CAD for all of its workload. All other organisations have to change direction gradually and with considerable care. Business direction is the concern of management. Redirecting a business towards the rapidly moving target of evolving new technology is something which can be both daunting and a challenge to any management.

Yet the signs are unmistakable. Planning and design work is a business activity. It must survive in a world which itself is changing and is highly competitive. Of course there is always the search for higher levels of productivity. Yet it seems that too much concentration on productivity is likely to blind us to a related but even greater issue – Design Quality.

The entire population is coming to expect better quality in all kinds of products and services. Owners, occupiers and users of construction projects as a group are unlikely to stand out as exceptions. Quality is reflected in many ways, such as fitness of purpose, and *zero* defects. Where quality is missing, sadly there is greater recourse today to the legal system.

This drive for quality presents special problems for an industry which is fragmented in its organisation, embraces multiple disciplines among its designers and planners, and which as a rule provides one-off solutions.

Up to now, drawing sheets have been individually crafted on a drawing board. Almost inevitably, the document set for a complete project will contain inconsistencies. So manual drawing production is unlikely to survive in the long-term. However to merely mimic the process using a computer is unlikely to provide the complete or fully effective solution.

This has led us to the concept of constructing prototypes in the form of computer models of projects. These are *soft* in the sense that they can exist within computer stores, but can be assessed and analysed in many ways, and viewed from all angles on display screens. The design is made *hard* only when we choose either to extract and use documentation from printers and plotters, or to transmit the data electronically to others. In itself, this use of computer models will not ensure that design solutions will be better and more easily constructable, but it presents better opportunities for those who wish to grasp them. Perhaps we will indeed begin to move away from the 'as-designed' and the 'as-built' syndrome.

We have seen that investment in CAD technology is one step, but a parallel move must be to set up the management structure, the training, and several operational and control procedures. The technology can be purchased, but the rest has to be organised by the design firm itself. Designing with computers is different. Retraining an individual to operate a workstation may not take too many days; retraining and reorganising the design office will take rather longer. Anyone who feels that it is all an unnecessary price perhaps should reflect on the time required for their profession training, and the overheads already involved in organising their work. CAD is not merely yet another computer program, it is basic to the whole design process.

Many design offices have a history of applying computers in various narrow areas of their work. In the past these application areas have been fragmented. They have been called 'islands of automation' in a sea of manual practice. A better description might be 'islands of excellence' for individually they are usually effective in their own limited spheres, and undertaken by well-meaning people.

The current trend however, and it appears to be gathering momentum, is towards greater integration of computer applications. The development of CAD is merely an important element within this trend. The concept of integration treats the design process as merely an early phase of a much more extensive process. This extends from the initial requirement, through conceptual and detail design, construction, commissioning, acceptance and payments, operation and maintenance, possibly to refurbishment activity, and onward until ultimate dismantlement.

Much of the information that would be of value to all the parties that contribute at various times is created during the design phase. In principle therefore it only seems logical for the design models to be maintained, added to and altered as appropriate, transmitted and made available to all that require them.

It is a grand scheme and I have only been able to deal with limited elements of it in this book. Management should be aware of the undercurrents however. If the total scheme appears too grand, then we should be aware of movements in other industries. CIM is an acronym for 'Computer Integrated Manufacture' and this is a concept of information transfer and integration of computer aided and computer controlled processes. It is being taken increasingly seriously in many influential quarters in industry. The building and construction industry has surprisingly similar problems and needs – at least in principle. Building and construction needs its own version of CIM.

References

1. Billington, Dr Colin and Latter, Dr Bob. 'Multi-storey Construction – Trends turn to Steel'. *Civil Engineering*. Oct 1986 pp23–6.
2. RIBA. (1983) *Architects' Job Book*. Fourth revised Edition.
3. Lawson, Bryan. (1980) *How Designers Think*. The Architectural Press Ltd. London.
4. Port, Dr Stanley. (1984) *Computer Aided Design for Construction. A Guide for Engineers, Architects and Draughtsmen*. Collins, London.
5. Port, Dr Stanley. (1985) Selection and Implementation of CAD for Small and Medium Firms. *Computer Graphics User '85 Conference*. London.
6. Port, S. (1986) 'Procedure for Selecting and Implementing CAD'. *Proceedings of CADCAM '86 Conference*, Birmingham. pp157–160.
7. Neesham, Claire. (1987) Consultants: 'Through the Maze'. *CADCAM International*. 6(8) pp31–41.
8. Port, Stanley. (1986) Do-it-yourself Graphics Packages. *CADCAM International*. 5(2), pp35–36.
9. CICA. (1985) *CAD Systems Evaluated for Construction*. Construction Industry Computing Association. Cambridge.
10. Woods, Tony. (1987) 'The Black Art of Benchmarking'. *CADCAM International*. 6(4) pp19–20.
11. Port, S and Myers, A.P. (1985) Computer Graphics and Reinforced Concrete Detailing. *Journal, The Institution of Structural Engineers*, 63A(1), pp15–17.
12. Lawrence, Andrew. (1987) Does CADCAM Pay? *CADCAM International*. 6(8) pp14–16.
13. Sharman, Andrew. 'The Making of Cities – What are we to Do? 1. The Challenge of Components, Constraints and Inspirations'. *Journal Royal Society of Arts* (1985) CXXXIII(5352) pp816–835.
14. Biss, M.A. and Arnot G.L. (1986) 'Living with CAD in Colchester Division, Anglian Water'. *CADCAM '86 Conference*. Birmingham. pp179–187.
15. Grindley, R.E. (1988) How Computer Networks Help Provide a Competitive Advantage. *Management of New Technology*. An Institution of Civil Engineers Conference. London. Thomas Telford. pp25–41.
16. Armer, D.M. (1988) Data Transmission Cables – 'The Necessary Evil'. *Management of New Technology*. An Institution of Civil Engineers Conference. London. Thomas Telford. pp1–10.
17. Holscher, David. Floor Plan: Need and Requirements for Good Data Centre Design. *Systems International* Oct 1986 pp96–99.

18. Jenkins, Alistair. (1987) Putting Pen to Paper. *Draughting & Design*. 7(7) pp14–15.
19. Port, Dr Stanley, and Tamplin, Peter. (1987) Are You Sitting Comfortably? *CADCAM International*. 6(8) pp23–24.
20. Jenkins, Alistair. (1987) Plotting Without Tears. *Draughting & Design*. 7(8) p7.
21. Tope, J.H. (1987) 'A Few Plotting Tips'. *Draughting & Design*. 7(7) p16.
22. Jones, Huw. (1987) 'Digital Map Revolution'. *New Civil Engineer*. 1–8 Jan 1987. pp22–23.
23. Smith, N. (1988) 'Ordnance Survey Digital Mapping'. *Management of New Technology*. An Institution of Civil Engineers Conference. London.Thomas Telford. pp91–103.
24. Bracewell, P.A. and Klement, U.R. (1983) 'The Use of Photogrammetry in Piping Design'. *The Institution of Mechanical Engineers*. 197A(30) 14pp.
25. Anon. (1986) Photometrology. *CICA Bulletin*. No 17. March 1986. p7.
26. Watkins, Geoffrey, W. (1985) Audio-visual Presentation Techniques using 3–D Computer Graphics. *Proceedings. Construction Industry Computer Conference 1985*. RIBA. London. 2pp.
27. t^2 Solutions Ltd. (1985) *The Esquisse*. A video cassette available from t^2 Solutions Ltd. The Teamwork Centre, Prince Edward Street, Berkhamsted, Herts HP4 3AY, England.
28. Herbert, Mark. (1987) 'A Walk on the Clydeside'. (Abacus). *CADCAM International*. 6(4) pp41–42.
29. Anon. (1985) 'Computer Animation Blasts-off'. *CADCAM International*. 4(10) p8.
30. Vince, John. (1987) 'The Real Thing. (Real-time computer graphics simulations at Rediffusion Simulations, Crawley)'. *CADCAM International*. 6(4) pp58–59.
31. Bridges, Dr Alan, H. (1985) Computer Visualisation of Architectural Designs. *Computer Graphics User '85 Conference*. London. 10pp.
32. Eastman, Charles M. (1986) 'Fundamental Problems in the Development of Computer-based Architectural Models'. *Symposium on Architectural CAD*. Massachusetts Institute of Technology & Graphic Systems Inc. July 4, 1986. pp8.
33. International Business Machines Corporation (IBM). (1982) *The Economic Value of Rapid Response Time*. IBM. Publication GE 20–0752–0 (11–82), Dept 824. 1133 Westchester Ave., White Plains, New York 10604. pp11.
34. MacDonald, Doug. (1987) 'Make CAD Manage the Database'. *CADCAM International*. 6(3) pp55–58.
35. Scott, George M. (1986) *Principles of Management Information Systems*. McGraw-Hill Book Company. New York.
36. Motor Columbus, Spie Balignolles, Socotec. (1984) *Quality Management Standard for Civil Works*. Macmillan Press. London.
37. Power, R.D. (1985) 'Quality Assurance in Civil Engineering'. *Construction Industry Research & Information Association Report 109*. London.
38. British Standards Institution. (1979) Quality Systems. Part 1: Specification for Design, Manufacture and Installation.
39. Harris, R.P. (1988) The Project Manager and Quality Assurance. *Management of New Technology*. An Institution of Civil Engineers Conference. London.Thomas Telford. pp117–29.
40. Port, Dr Stanley. (1987) *An ABC of CAD. Interior Designers Handbook – 1987*. Grosvenor Press International Ltd. London. pp316–19.

Index